Fundamentals of Infrared Sensing

Tutorial Texts Series Related Title

- *Infrared Design Examples*, W. L. Wolfe, Vol. TT36
- *Infrared Optics and Zoom Lenses, Second Edition*, A. Mann, Vol. TT83
- *Optical Design Fundamentals for Infrared Systems, Second Edition*, Max J. Riedl, Vol. TT 48
- *Practical Applications of Infrared Thermal Sensing and Imaging Equipment, Third Edition*, H. Kaplan, Vol. TT75

(For a complete list of Tutorial Texts, see http://spie.org/publications/books/tutorial-texts.)

Other related SPIE Press books:

- *Field Guide to Infrared Systems, Detectors, and FPAs, Third Edition*, A. Daniels, Vol. FG40
- *Field Guide to Photographic Science*, A. D. Rowlands, Vol. FG46
- *Materials for Infrared Windows and Domes: Properties and Performance, Second Edition*, D. C. Harris, Vol. PM368
- *Short-Wavelength Infrared Windows for Biomedical Applications*, L. A. Sordillo and P. P. Sordillo, Eds., PM336
- *Testing and Evaluation of Infrared Imaging Systems, Third Edition*, G. C. Holst, Vol. PM185

Fundamentals of Infrared Sensing

Glenn D. Boreman

Tutorial Texts in Optical Engineering
Volume TT128

Jessica DeGroote Nelson, Series Editor

SPIE PRESS
Bellingham, Washington USA

Library of Congress Cataloging-in-Publication Data

Names: Boreman, G. D. (Glenn D.) author
Title: Fundamentals of infrared sensing / Glenn D. Boreman.
Description: Bellingham, Washington, USA : SPIE Press, [2025] | Series: SPIE tutorial texts in optical engineering ; volume TT128 | Includes bibliographical references and index.
Identifiers: LCCN 2024036602 | ISBN 9781510682337 paperback | ISBN 9781510682344 pdf | ISBN 9781510682351 epub
Subjects: LCSH: Infrared detectors
Classification: LCC TA1573 .B67 2025 | DDC 621.36/2–dc23/eng/20250616
LC record available at https://lccn.loc.gov/2024036602

Published by

SPIE
P.O. Box 10
Bellingham, Washington 98227-0010 USA
Phone: +1 360.676.3290
Fax: +1 360.647.1445
Email: books@spie.org
Web: www.spie.org

The content of this book reflects the work and thought of the author. Every effort has been made to publish reliable and accurate information herein, but the publisher is not responsible for the validity of the information or for any outcomes resulting from reliance thereon.

Printed in the United States of America.
First printing 2025.
For updates to this book, visit http://spie.org and type "TT128" in the search field.

SPIE.

Introduction to the Series

The *Tutorial Text* series provides readers with an introductory reference text to a particular field or technology. The books in the series are different from other technical monographs and textbooks in the manner in which the material is presented. True to their name, they are tutorial in nature, and graphical and illustrative material is used whenever possible to better explain basic and more-advanced topics. Heavy use of tabular reference data and numerous examples further explain the presented concept. A grasp of the material can be deepened and clarified by taking corresponding SPIE short courses.

The initial concept for the series came from Jim Harrington (1942–2018) in 1989. Jim served as Series Editor from its inception to 2018. The *Tutorial Texts* have grown in popularity and scope of material covered since 1989. They are popular because they provide a ready reference for those wishing to learn about emerging technologies or the latest information within a new field. The topics in the series have grown from geometrical optics, optical detectors, and image processing to include the emerging fields of nanotechnology, biomedical optics, engineered materials, data processing, and laser technologies. Authors contributing to the series are instructed to provide introductory material so that those new to the field may use the book as a starting point to get a basic grasp of the material.

The publishing time for *Tutorial Texts* is kept to a minimum so that the books can be as timely and up-to-date as possible. When a proposal for a text is received, it is evaluated to determine the relevance of the proposed topic. This initial reviewing process helps authors identify additional material or changes in approach early in the writing process, which results in a stronger book. Once a manuscript is completed, it is peer reviewed by multiple experts in the field to ensure that it accurately communicates the key components of the science and technologies in a tutorial style.

It is my goal to continue to maintain the style and quality of books in the series and to further expand the topic areas to include new emerging fields as they become of interest to our readers.

Jessica DeGroote Nelson
Edmund Optics

Contents

Preface *xi*

1 Introduction **1**

References 5

2 Geometrical Optics **7**

2.1 Introduction 7

2.2 Imaging Concepts 7

2.3 Algebraic Rules for Image Formation 8

2.3.1 Point sources at infinity 10

2.4 Aperture Stop, Marginal Ray, and F-number 11

2.5 Field of View 12

2.6 Detector Footprint 13

2.7 Image Quality 14

2.7.1 Diffraction 14

2.7.2 Aberrations 16

2.7.3 Image-quality summary 16

2.8 Infrared Materials 17

2.9 Modulation Transfer Function 19

2.9.1 Spatial frequency 20

2.9.2 Modulation transfer function definition 21

2.9.3 MTF examples 25

References 26

Bibliography 26

3 Radiometry **27**

3.1 Introduction 27

3.1.1 Solid angle 27

3.1.2 Radiometric nomenclature and units 29

3.2 Irradiance and Exitance 29

3.3 Intensity 31

3.3.1 Inverse-square law for point sources 32

3.3.2 Irradiance on a tilted receiver 32

3.4 Radiance 33

3.4.1 Radiance example for a tilted receiver 36

3.4.2 Lambertian radiator 36
3.4.3 Nonzero view angles 37
3.4.4 Cosine-to-the-fourth falloff 37
3.4.5 Tilting a Lambertian source 38
3.4.6 Lambertian radiator: relation of exitance and radiance 39
3.5 Flux Transfer in Image-Forming Systems 40
3.5.1 Why is radiance analogous to perceived brightness? 41
3.5.2 Example calculation of image irradiance (extended source) 42
3.5.3 When to use a lens? 43
3.5.4 Cosine-to-the-fourth falloff in imaging systems 44
3.5.5 Irradiance depends on the image-space F-number 45
3.5.6 Aperture stop not at the lens 46
3.5.7 Radiance of an extended source is conserved 47
3.5.8 Dilution of radiance by a finite impulse response 48
3.6 Point-Source Imaging 48
Bibliography 49

4 Thermal Sources 51

4.1 Introduction 51
4.2 Blackbody Radiation 52
4.3 Planck's Equation 54
4.3.1 Other forms of Planck's equation 54
4.3.2 Stefan–Boltzmann law 55
4.3.3 Wien's displacement law 56
4.3.4 Half-power points 57
4.3.5 Calculations 58
4.3.6 Planck's equation in terms of other spectral variables 59
4.4 Exitance Contrast 59
4.5 Emissivity 60
4.5.1 Kirchhoff's law 61
4.5.2 Graybody 62
4.5.3 Selective radiator 63
4.5.4 Emissivity of common materials 63
4.5.5 Emissivity as a function of λ and T 63
References 64
Bibliography 65

5 Detectors 67

5.1 Comparing Thermal Detectors and Photon Detectors 67
5.2 Responsivity 69
5.2.1 Spectral responsivity 70
5.2.2 Blackbody responsivity 76
5.2.3 Relating $\mathcal{R}(T)$ to $\mathcal{R}(\lambda_{cut})$ 76

5.3 Cooling Requirements for Photon Detectors 78
5.4 Time Response and Frequency Response 79
5.5 Thermal Detectors 81
5.5.1 Temporal response 81
5.5.2 Bolometric sensors 83
5.5.3 Pyroelectric sensors 85
5.6 Photon Sensors 89
5.6.1 Photoconductive sensors 89
5.6.1.1 Intrinsic vs extrinsic PC 89
5.6.1.2 Photoconductive mechanism 91
5.6.2 Photovoltaic mechanism 92
5.6.2.1 Open-circuit mode 94
5.6.2.2 Short-circuit mode 94
5.6.2.3 Reverse-bias mode 95
5.6.3 Schottky-barrier detectors 96
Reference 97
Bibliography 97

6 Temperature Measurement 99
6.1 Measurement Configurations 99
6.2 Radiation Temperature 101
6.3 Brightness Temperature 101
6.4 Color Temperature 103
6.5 Practical Temperature Measurements with an IR Camera 104
Bibliography 106

7 Noise in the Detection Process 107
7.1 Basic Noise Nomenclature 107
7.2 Sources of Noise 108
7.3 Internally Generated Sensor Noise 109
7.4 Noise Power Spectral Density 110
7.5 Calculations with White Noise 112
7.6 Shot Noise 113
7.6.1 Development of the shot-noise expression 114
7.6.2 What process generates the current? 115
7.7 SNR in the Signal-Shot-Noise Limit 116
7.8 SNR in the Background-Shot-Noise Limit 119
7.9 Generation–Recombination Noise 121
7.10 Johnson Noise 121
7.10.1 Johnson-noise units 122
7.10.2 Johnson-noise calculation example: two parallel resistors at different temperatures 123
7.10.3 SNR in the Johnson-noise limit 123
7.10.4 Johnson noise root-area and root-bandwidth dependence 123

7.11 1/f Noise 124
7.12 Thermal-Fluctuation Noise 125
References 126
Bibliography 126

8 Detector Sensitivity Figures of Merit 127
8.1 Noise-Equivalent Power (NEP) 127
8.1.1 Numerical-calculation example with NEP 129
8.2 Normalized Detectivity D^* 129
8.2.1 Numerical calculations with D^* 131
8.2.2 Measurement procedure for D^* 132
8.3 Background-Limited D^* for Photon Sensors 133
8.4 Johnson-Noise-Limited (JOLI) D^* for Photon Sensors 137
8.5 Temperature-Fluctuation-Noise-Limited D^* for Thermal Sensors 139
8.6 Comparing the Performance of Thermal and Photon Sensors 140
Reference 141
Bibliography 141

9 Infrared Systems 143
9.1 General Comments 143
9.2 Scan Formats 143
9.2.1 Multiple-detector scan formats 146
9.3 Search System: Range Equation 147
9.3.1 Example of a search system preliminary design 150
9.3.1.1 Finding the maximum frame rate consistent with initial specs 150
9.3.1.2 Finding the maximum frame rate using a one-dimensional detector array 151
9.3.1.3 Using excess SNR to increase the maximum range 152
9.3.1.4 Using excess SNR to decrease the front-end size 152
9.4 Thermal Imager: NETD Expression 152
9.4.1 Calculation example for NETD 155
9.5 Minimum Resolvable Temperature Difference (MRTD) 155
References 157
Bibliography 158

Index 159

Preface

This book is an outgrowth of a short course that I have presented for SPIE for the past 25 years or so on the fundamentals of infrared sensing. This field spans several technical disciplines, which can leave the beginner with the "where do I start?" question. My selection of the material included here represents those concepts and terminology that a newcomer in the field of infrared systems needs to understand. These include items that were initially confusing to me, ones that I have found useful in practice, or concepts that are commonly misunderstood. I have tried to explain these in the simplest terms that keep the key ideas. The level of mathematics is generally algebra-based. The book contains sample calculations but does not include problem sets. This is a consequence of the book's origin as a set of short course notes. It is not intended as a stand-alone college textbook but rather a self-study reference for the beginning systems engineer. The intended audience is a person with a bachelor's-level training in science or engineering. It is meant to be an introduction, with sufficient detail to enable the reader to make initial back-of-the-envelope calculations and to understand the basic tradeoffs and trends involved.

During the writing process, I appreciated daily encouragement via WhatsApp from Prof. Mike Sundheimer of the Universidade Federal Rural de Pernambuco in Recife, Brazil, weekly encouragement via Zoom from Skye Engel, as well as facilities support and encouragement from Prof. Ian Ferguson of Kennesaw State University. I acknowledge helpful discussions with Dr. Nenad Vrucinic. I am grateful to Dr. Austin Richards for providing several infrared images, to various sources for the permissions granted for figure reproductions, to the anonymous reviewers for their insightful and helpful comments, and to Dara Burrows, Senior Editor at SPIE Press. Last, but surely not least, I want to thank Maggie Boreman—my wife of 35 years, my main encourager and technical editor, who has once again shouldered the task of transforming my writing into readable English.

Glenn D. Boreman
Emerald Rose Farm
May 2025

Chapter 1
Introduction

Objects in the temperature range between 300 Kelvin (room temperature) and 1000 Kelvin radiate their energy primarily in the infrared (IR) portion of the spectrum. A scene can be visualized by means of its emitted radiation rather than by reflected radiation from the sun or from artificial sources.

Infrared imaging systems are capable of better angular resolution than radio-frequency systems because of the comparatively smaller wavelength. Visible-band imaging systems have better resolution although their operation is limited to clear weather and daylight conditions. A major challenge of IR systems is that the background is typically of similar temperature to the target of interest—thus the background also emits radiation. This means that IR images are inherently of lower contrast than comparable visible-band images, and thus we are always concerned with the signal-to-noise ratio (SNR) in the IR to maximize the detectability of image features. To make the infrared scene visible, an IR imaging system depends on the interaction of the several subsystems seen in Fig. 1.1: source, atmosphere, optics, detectors, electronics, image processing, display, and human observer.

We will consider the following questions:

- How do the temperature and surface characteristics of the source determine the wavelength and power distribution of the emitted radiation?
- What is the resolution of the imaging system?
- What is the field of view?
- How much power is collected from the source onto a given detector?
- By what mechanism is the power collected by the optics turned into an electrical signal?
- What is the wavelength response of the detectors?
- What voltage level is expected at the output of the detectors?
- What are the sources of noise?
- What is the post-detection signal-to-noise ratio?
- What electrical bandwidth is required to pass the signal data?

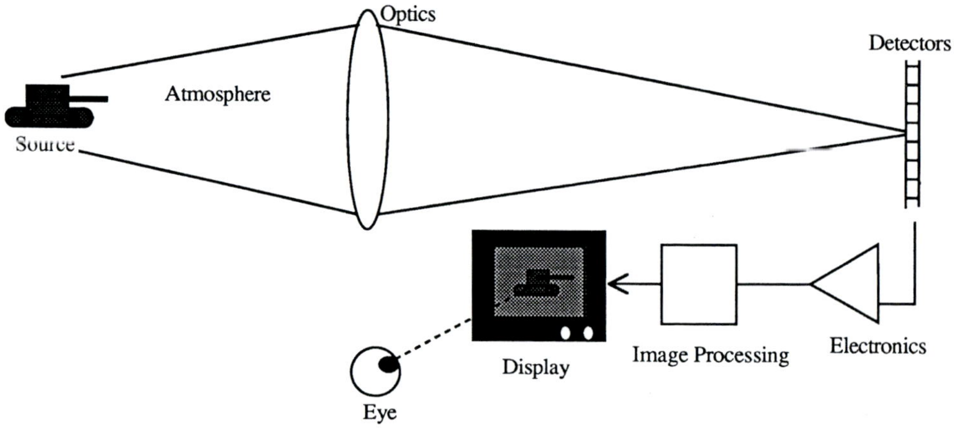

Figure 1.1 Infrared imaging involves interaction of several subsystems.

Historically, IR systems have received substantial developmental resources because of well-known applications in the defense sector. One byproduct of this long-term research-and-development effort has been a significant cost reduction in IR focal-plane arrays. This cost reduction has opened IR imaging applications for which the required components were previously too expensive for the applications to be commercially viable. An expanded range of applications led to further cost reductions because of economies of scale in production, leading to wider adoption of the technology. IR imaging is used to visualize heat sources in a variety of contexts.

The first wide-scale application of IR imaging was in defense sensing, as seen in Fig. 1.2. The hot engine compartment on airborne or terrestrial vehicles emits a significant amount of IR radiation.

Figure 1.2 Infrared image of helicopters in flight. (Image courtesy of Teledyne FLIR, LLC.)

Figure 1.3 Infrared imaging of buildings can identify heat-loss regions. (Image courtesy of Infrared Imaging Services, LLC.)

As seen in Fig. 1.3, IR images provide a map of temperature variations on the exterior surfaces of a building. This may allow identification of heat-loss and air-leakage locations in building envelopes, which helps to determine where a building needs additional insulation. Because wet insulation conducts heat better than dry insulation, thermographic scans can also detect roof leaks.

As seen in Fig. 1.4, IR images can be used to show loose electrical connections, which cannot be seen with the naked eye, before the additional resistance and consequent heating cause a fire or an electrical failure. IR images can also confirm expected temperature distributions in microcircuits.

As seen in Fig. 1.5, IR imaging can be used to visualize the location of heat pipes installed in a floor and to identify faulty connections.

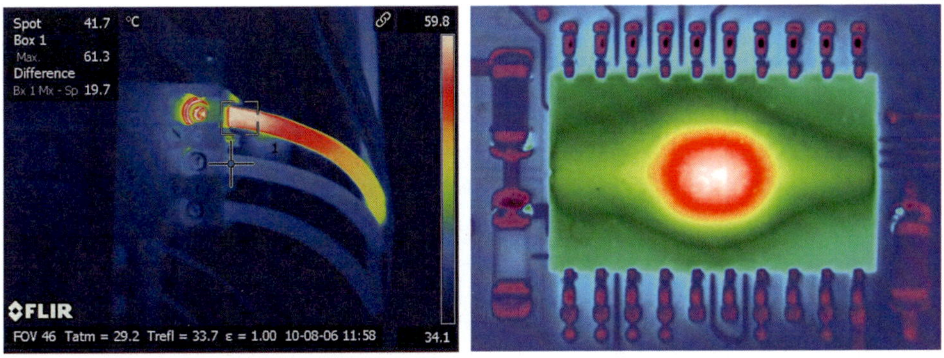

Figure 1.4 Infrared imaging can (left) identify loose electrical connections or (right) confirm proper circuit operation. (Left image courtesy of Infrared Imaging Services, LLC. Right image courtesy of Teledyne FLIR, LLC.)

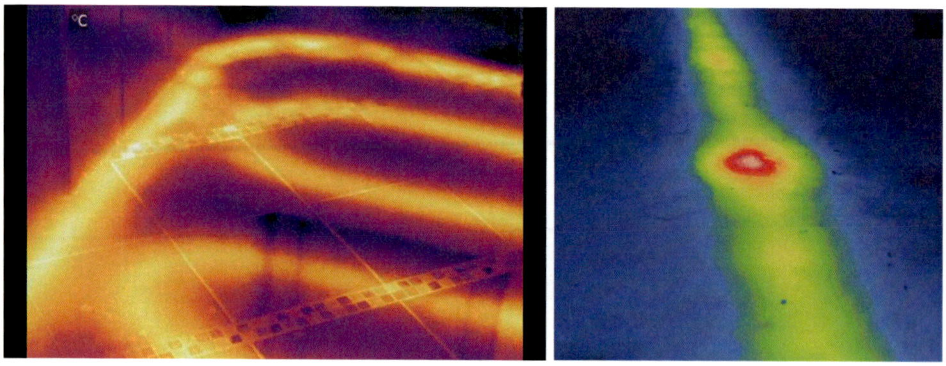

Figure 1.5 Infrared imaging can locate heat pipes under floors. (Images courtesy of Infrared Imaging Services LLC.)

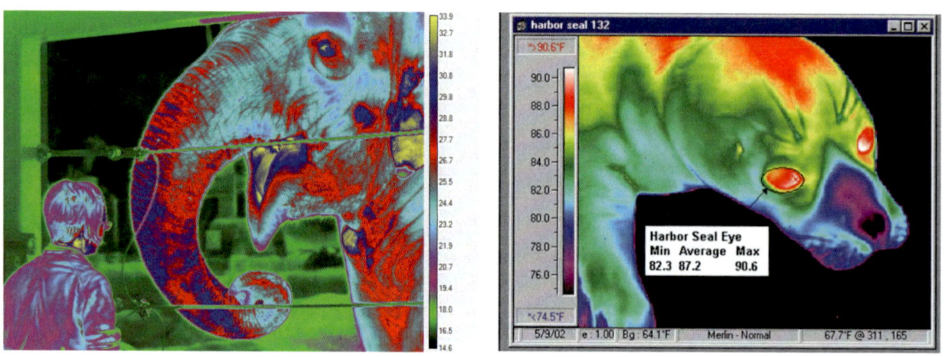

Figure 1.6 Infrared imaging can highlight regions of interest for veterinary examination. (Images courtesy of Teledyne FLIR, LLC.)

There is a high degree of bilateral symmetry in a normal animal or human body. By looking for anomalies in the thermal images seen in Fig. 1.6, one can infer locations of injury on animals which cannot be seen by visual inspection.

There are myriad applications in the medical treatment of humans, using temperature variation on the skin to visualize blood perfusion in injured tissue, circulation issues, pinched nerves, and other conditions, as seen in Fig. 1.7.

Infrared imaging can also be used to provide data for control of industrial processes that are temperature dependent or for status monitoring of critical parts, as seen in Fig. 1.8.

Some important topics related to IR sensing are not covered in this book. Because of the rapid pace of technological innovation, there continue to be significant developments in IR detectors, such as blocked-impurity-band IR detectors, superlattice-based sensors, and high-pixel-count IR focal-plane arrays. Along with Refs. 1–4, the *Proceedings of the Infrared Technology and Applications* conference, now held as part of the annual SPIE Defense and Commercial Sensing symposium, are useful for staying informed on the latest advances.

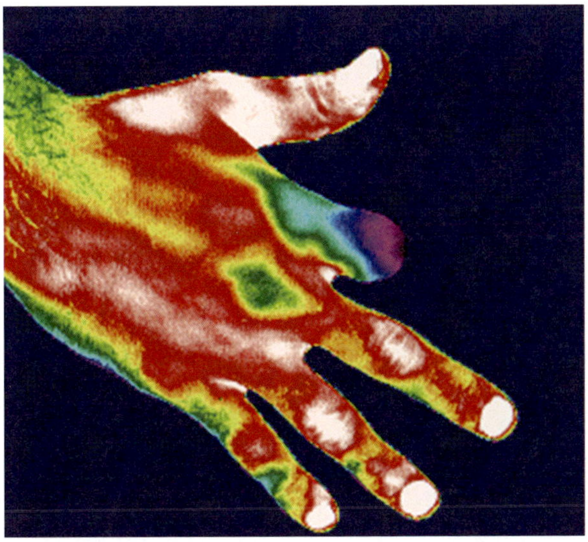

Figure 1.7 Infrared imaging can visualize blood-flow anomalies. (Image courtesy of Teledyne FLIR, LLC.)

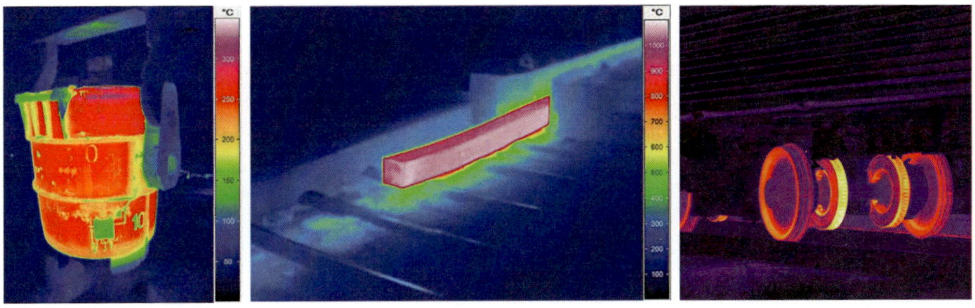

Figure 1.8 Infrared imaging provides data for manufacturing-process control, or for in-situ monitoring of critical components. (Left and center images courtesy of InfraTec GmbH Infrarotsensorik und Messtechnik (www.InfraTec.de). Right image courtesy of Teledyne FLIR, LLC.)

References 5 and 6 are useful resources for a more detailed treatment of the semiconductor-physics principles behind the sensing modalities described. References 7 and 8 are good sources of information on the image-processing principles that are commonly employed in IR imaging applications, and Ref. 9 is a starting point for understanding the read-out integrated circuits (ROICs) that are a critical component of a modern-day IR imager system.

References

1. M. A. Kinch, *State-of-the-Art Infrared Detector Technology*, SPIE Press, Bellingham, Washington (2014) [doi: 10.1117/3.1002766].

2. A. Rogalski, M. Kopytko, and P. Martyniuk, *Antimonide-based Infrared Detectors: A New Perspective*, SPIE Press, Bellingham, Washington (2018) [doi: 10.1117/3.2278814].
3. A. Rogalski, K. Adamiec, and J. Rutkowski, *Narrow-Gap Semiconductor Photodiodes*, SPIE Press, Bellingham, Washington (2000) [doi: 10.1117/3.2536930].
4. M. A. Kinch, *Fundamentals of Infrared Detector Materials*, SPIE Press, Bellingham, Washington (2007) [doi: 10.1117/3.741688].
5. E. Martins, *Essentials of Semiconductor Device Physics*, Wiley, New York (2022).
6. S. Sze, Y. Li, and K. Ng, *Physics of Semiconductor Devices, 4th Edition*, Wiley, New York (2021).
7. A. R. Weeks, *Fundamentals of Electronic Image Processing*, SPIE Press, Bellingham, Washington (1996) [doi: 10.1117/3.227778].
8. R. Fiete, *Modeling the Imaging Chain of Digital Cameras*, SPIE Press, Bellingham, Washington (2010) [doi: 10.1117/3.2184893].
9. K. Veeder, *Digital Converters for Image Sensors*, SPIE Press, Bellingham, Washington (2015) [doi: 10.1117/3.1000639].

Chapter 2
Geometrical Optics

2.1 Introduction

In this chapter we introduce the concepts from geometrical optics needed to understand the operation of IR imaging systems. From the perspective of an optical designer, more detailed analyses including computer modeling are necessary to accomplish satisfaction of specifications for both image quality and efficiency of flux throughput. We begin with a brief overview of imaging concepts, define the parameters F-number (*F*/#) and field of view (FOV), and consider issues that affect image quality. Then, we consider the spectral characteristics of several materials that are useful in the IR band for transmissive and reflective elements. We finish this chapter with a brief consideration of an alternative means of describing and specifying image quality: the modulation transfer function (MTF).

2.2 Imaging Concepts

We begin our discussion by defining what is meant by objects and images. An object is a collection of independent source points contained in a plane, each with a different strength. Light rays are lines of power transport away from the point sources. Each point source emits rays into all forward directions (that is, toward the imaging system). The spatial distribution of brightness across the object plane is the information we want to reproduce in the image plane. Typically, a planar detector array located at the image plane produces a matrix of voltage outputs related to the local spatial power density in the image plane. This power density is denoted as irradiance, which is specified in units of W/cm^2.

We can understand this process by referring to Fig. 2.1, where we visualize rays arising from both on- and off-axis object-plane point sources. These rays pass through the glass lens, which changes the direction of the rays because of refraction at the air/glass interfaces. The rays then propagate to their corresponding image-plane intersections. Rays that diverge from any object point will intersect at their corresponding image-plane location. The image is

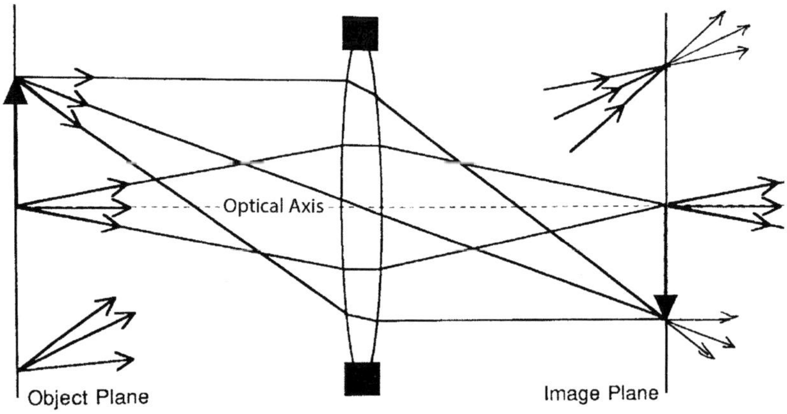

Figure 2.1 Image formation.

thus formed on a point-by-point basis. The power density at any image location is proportional to the brightness of a corresponding object point. The result is a geometrically similar distribution of brightness. In optical diagrams, an object/image relationship occurs when rays diverge from one point and re-intersect at another location.

In Fig. 2.1 the optical axis is seen to be an axis of rotational symmetry for the optical system, containing the centers of curvature of the lens surfaces, which are portions of spheres. The optical axis is also called the line of sight because it defines the direction that the optical system is looking. The optical axis generally defines the center of the scene being imaged because typically the object and image are centered on the optical axis. The object and image planes are assumed to be perpendicular to the optical axis. From an image-quality perspective, the fidelity of the image (that is, how similar it is to the object) is best near the optical axis. Also, in that figure we note a graphical-representation convention, where a single-headed arrow from the optical axis to the top of the object tracks the orientation of the image.

2.3 Algebraic Rules for Image Formation

The focal length of the lens f is positive for a lens that focuses an incoming beam of parallel light rays. A lens that is thicker in the middle than at the periphery has a positive f. A focal point exists on each side of the lens, as seen in Fig. 2.2.

The relevant variables seen in Fig. 2.3 are defined as follows, along with the sign convention for each. The object-to-lens distance p is positive to the left. The lens-to-image distance q is positive to the right. The object height and image height, h_{obj} and h_{img}, respectively, are measured from the optical axis to the edge of the region being imaged. Distances above the axis are positive and distances below the axis are negative. Because the object and image are

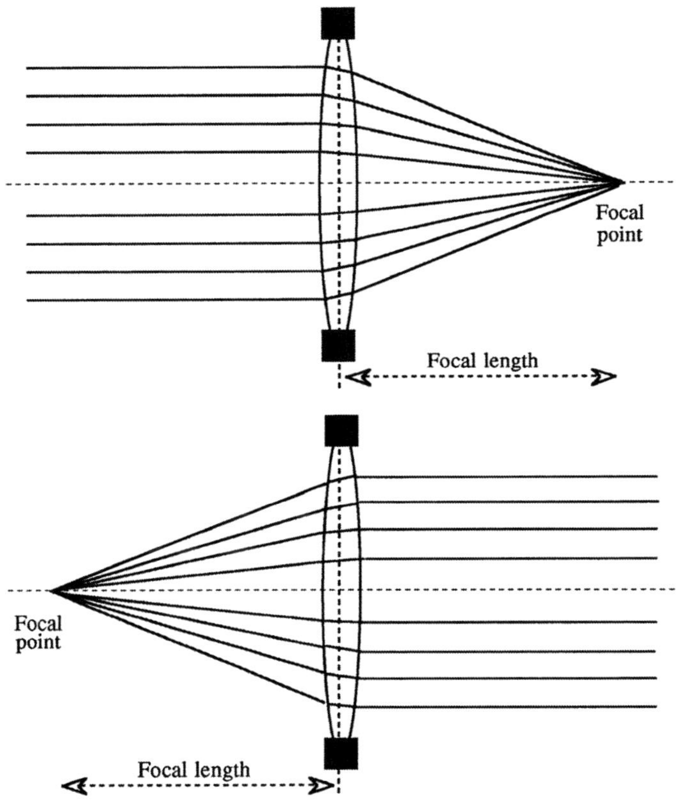

Figure 2.2 Focal length.

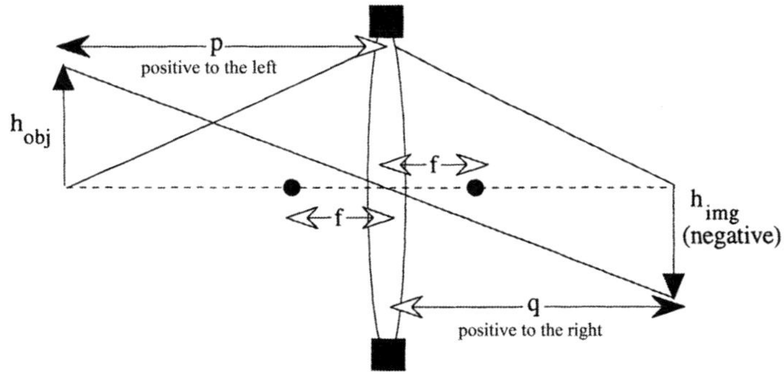

Figure 2.3 Variable definitions for imaging equation.

assumed to be centered on the optical axis, h_{obj} and h_{img} are *half-heights*. Only the off-axis distance needs be specified because of the assumed rotational symmetry of the lens about the optical axis.

Under this sign convention, the quantities p, q, f, and h_{obj} in Fig. 2.3 are positive, and h_{img} is negative. If we know the distance p from the object plane to the lens as well as the focal length f of the lens, we can determine the

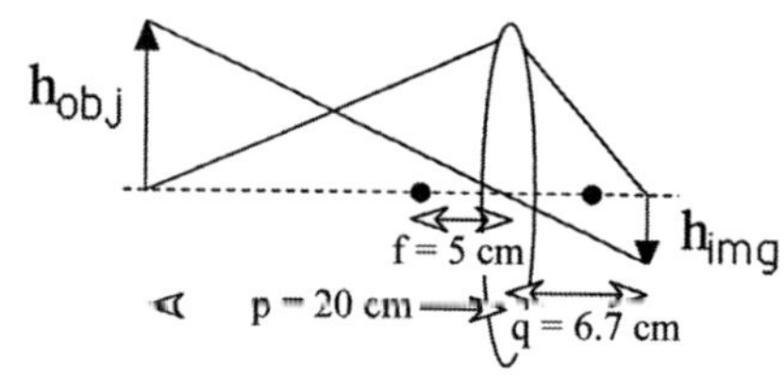

Figure 2.4 Example use of the imaging equations.

lens-to-image-plane distance using Eq. (2.1a) and the image magnification using Eq. (2.1b):

$$\frac{1}{p}+\frac{1}{q}=\frac{1}{f} \tag{2.1a}$$

$$\text{Magnification } \mathcal{M} \equiv \frac{h_{\text{img}}}{h_{\text{obj}}} = -\frac{q}{p}. \tag{2.1b}$$

These are approximate solutions, under the assumptions that the axial thickness of the lens is small compared to the focal length and that the angles that rays make with the optical axis are also small (so-called *paraxial* rays). Equations (2.1a) and (2.1b) are a convenient starting point for analysis, and this level of accuracy will suffice for the systems-level discussions considered in this book. As an example (Fig. 2.4), if we are given $f = 5$ cm and $p = 20$ cm, we find that $q = 6.67$ cm and $\mathcal{M} = -0.33$.

From Eq. (2.1a) we see that as $p \to \infty$, $q \to f$, consistent with the upper diagram of Fig. 2.2. As an object gets farther away, the image distance gets closer to f. Similarly, as $p \to f$, we find that $q \to \infty$. As an object is placed closer to the front focal point, the image distance gets farther away, as shown by the lower diagram in Fig. 2.2. A three-dimensional scene can be analyzed considering one object plane at a time. From Eq. (2.1a) we see that a wide range of distances in the object plane are mapped to a very small range of distances in the image plane. For instance, if the object is 100 m from a lens with a 40-mm focal length, $1/100 + 1/q = 1/(40 \times 10^{-3})$ yields an image distance $p = 40.016$ mm. This image distance is just a little more than the focal length. If the object is 200 m away, then $1/200 + 1/q = 1/(40 \times 10^{-3})$ yields a new image distance $p = 40.008$ mm. When the object is farther away, the image is 8 μm closer to the focal point. If, for a given range of object distances, these image locations are sufficiently close that negligible defocusing occurs, we can consider there to be a single image plane. For distant objects, the image plane is approximately one focal length from the lens. This is the usual situation for IR imaging systems in actual use, and we typically make the approximation $q \approx f$.

2.3.1 Point sources at infinity

Point sources a long distance away from the optical system give rise to parallel bundles of rays, as seen in Fig. 2.5. For an on-axis point at infinity,

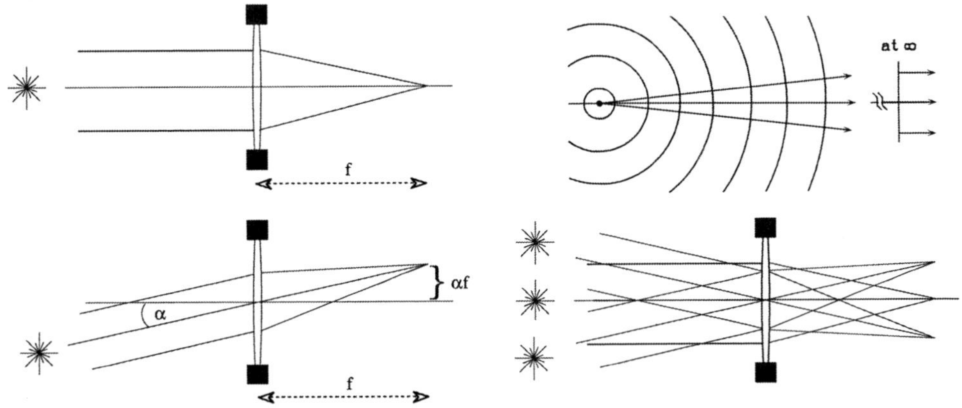

Figure 2.5 Constructions for point sources at infinity.

the parallel bundle of rays focuses on-axis, at f behind the lens, according to Eq. (2.1a). For a point at infinity located at a small angle α [radians] off-axis, the parallel bundle of rays focuses at a distance f behind the lens, at a distance αf from the axis. Angles in object space are thus mapped to locations in the image plane.

2.4 Aperture Stop, Marginal Ray, and F-number

We next consider how to specify how much aperture the system uses. We define an axial ray as one that starts at the on-axis point in the object and ends at the on-axis point in the image. As seen in Fig. 2.6, the aperture stop (ASTOP) is the surface that limits the angle over which the lens will accept rays from the on-axis object point. The marginal ray is a special axial ray that starts at the on-axis point in the object, goes through the edge of the aperture stop, and ends at the on-axis point in the image. The marginal ray is used to define the *F/#*.

The *F/#* is a parameter used to describe the ratio between the focal length f, the object distance p or image distance q, and the diameter D of the aperture stop. For an object at infinity, the image-space *F/#* is f/D, and for an object not at infinity, an *F/#* exists for both object space and image space, as seen in Fig. 2.7.

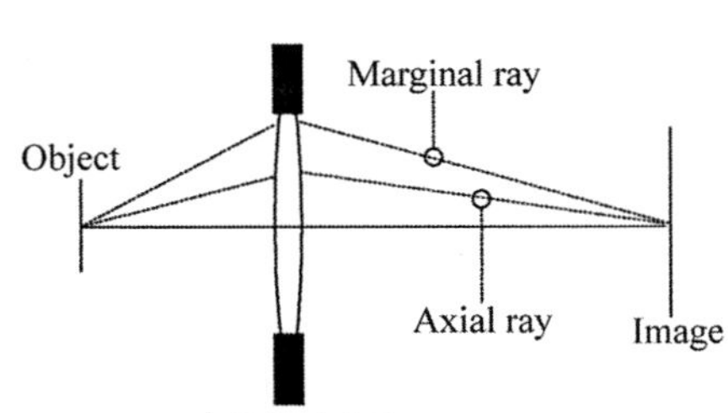

Figure 2.6 Aperture stop and marginal ray.

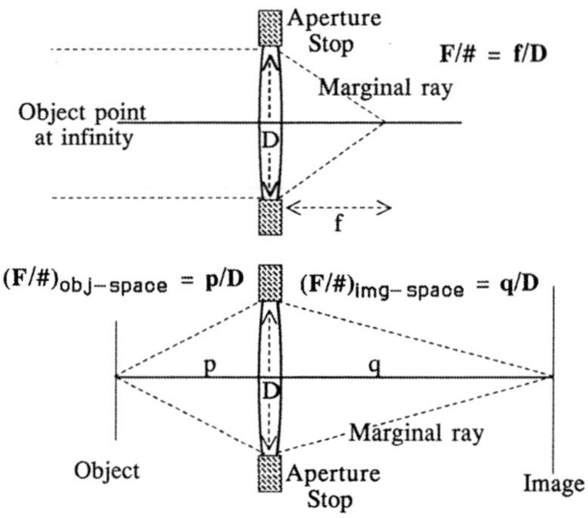

Figure 2.7 F-number is defined for both an object at infinity and for an object at a finite distance.

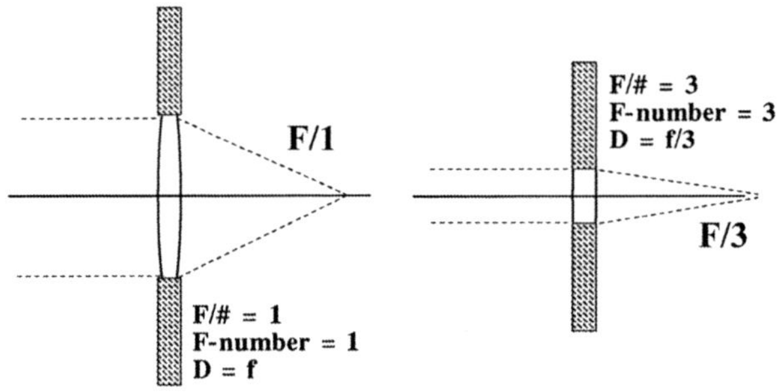

Figure 2.8 F-number decreases with increasing aperture diameter.

In the usual situation where the focal length f is fixed and the aperture-stop diameter can be adjusted, we note that a larger D gives a smaller $F/\#$, and a smaller D gives a bigger $F/\#$, as seen in Fig. 2.8.

2.5 Field of View

The field of view (FOV) is the angular coverage of an optical system. The FOV can be expressed as either a full angle or half angle, as seen in Fig. 2.9. Be aware that end users and systems engineers are usually concerned with full-angle FOV when considering the question of how much of the object scene can be imaged by the system. However, optical designers tend to think in terms of half-angle FOV (denoted here as FOV/2). The assumed rotational symmetry of the system about the optical axis makes it such that the designer only needs to consider off-axis heights in one direction in their design tasks. The FOV is physically determined by the size of the image-recording medium (usually a detector array; but it can also be a single detector or a piece of photographic film). The surface that limits the FOV is called the field stop. Limiting the field

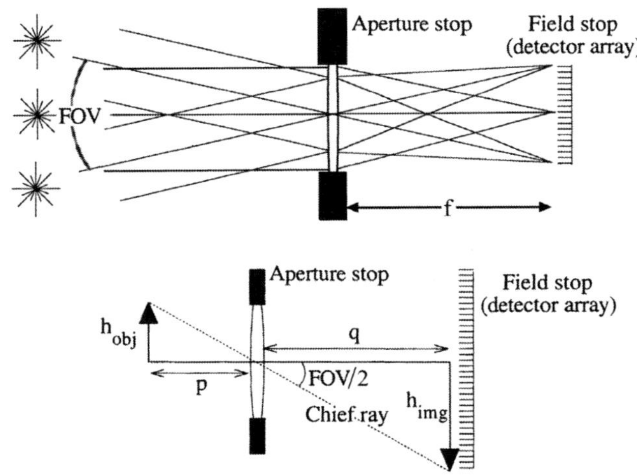

Figure 2.9 FOV for an object at infinity and for a near object.

coverage of an optical system helps to control image-quality defects because the image fidelity is usually best for image locations closer to the optical axis.

The chief ray goes from the maximum extent of the object to the maximum extent of the image, through the center of the aperture stop. The half-angle FOV is the angle that the chief ray makes with the optical axis.

2.6 Detector Footprint

A single detector located in the image plane has a *footprint*, also called the instantaneous field of view (IFOV), which is the image of the detector at the object plane. The IFOV defines the region of the object that contributes flux onto that sensor, as seen in Fig. 2.10, where we assume that object distance p is large enough that the image is formed at f behind the lens.

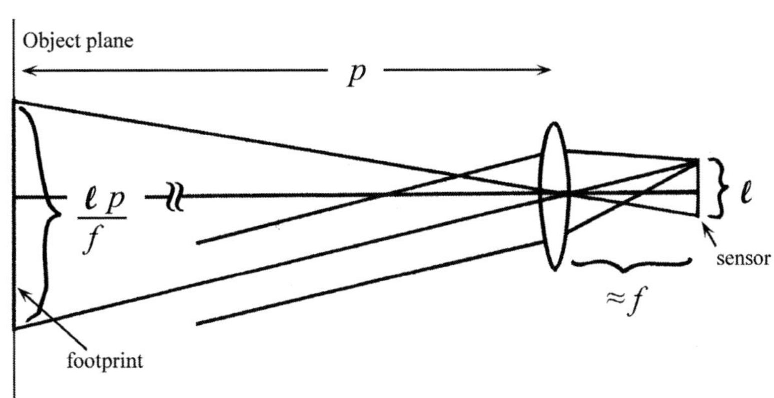

Figure 2.10 Detector footprint.

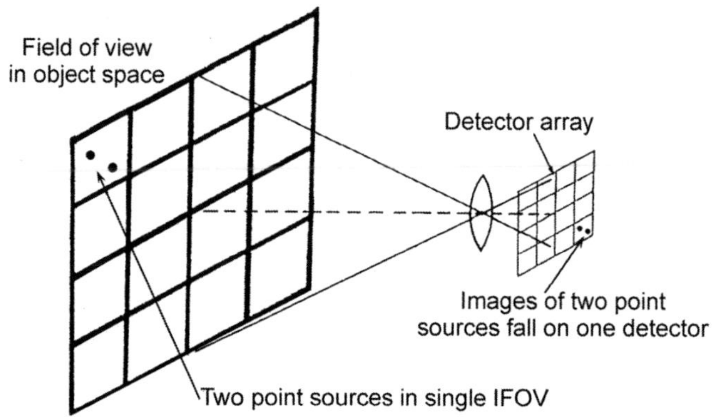

Figure 2.11 Irradiance is averaged across a pixel.

Assuming perfect imaging according to Eqs. (2.1a) and (2.1b), the detector footprint is a good estimate of the resolution element of the imaging system, referred to the object plane. As seen in Fig. 2.11, whatever power falls on an individual detector produces a single sensor output; that is, there is no spatial resolution within the IFOV.

2.7 Image Quality

Thus far we have assumed that point sources in the object are imaged to point locations in the image. However, because of the effects of diffraction, aberrations, and defocus, points in the object are mapped to blur spots (rather than points) in the image. Better image quality is obtained for smaller blur spots.

2.7.1 Diffraction

Because of the wave nature of light, an optical system with a finite-sized aperture stop can never form a point image. A characteristic minimum blur diameter forms in the image plane, even when other image defects are absent. The best image quality that we can achieve is determined by diffraction effects. The diameter of the diffraction blur in the image plane can be expressed by Eq. (2.2) as

$$d_{\text{diffraction}} = 2.4\ \lambda(F/\#)_{\text{img-space}}, \tag{2.2}$$

as shown in Fig. 2.12. The diameter of the diffraction blur is thus directly proportional to the operating wavelength of the system.

The normalized irradiance in the diffraction image of a point source as a function of distance ρ from the center of the pattern can be expressed in terms of a Bessel function of the first kind as $|2J_1(\pi\rho)/(\pi\rho)|^2$, as seen in the

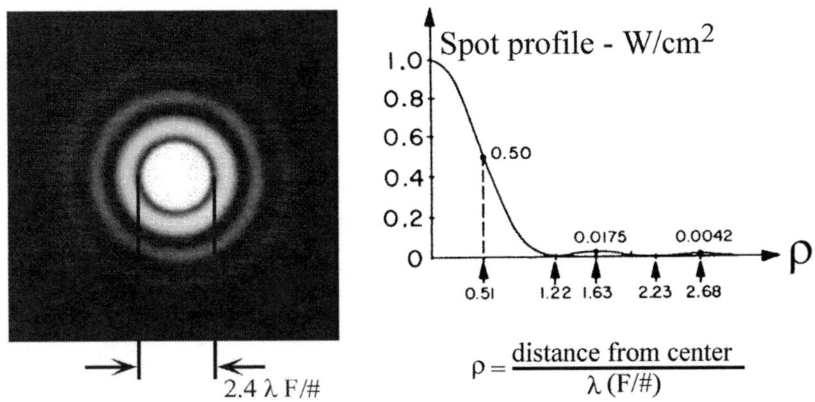

Figure 2.12 Dimension of the diffraction blur spot depends on wavelength and *F*/#.

right-hand diagram in Fig. 2.12. A circular aperture stop has been assumed. The variable ρ is the distance from the center of the irradiance distribution, normalized with respect to wavelength and *F*/#. The diffraction image has a ring structure. Equation (2.2) is the usual definition of diffraction spot size as used by the optics community, which is the diameter to the first null of the irradiance pattern. This diameter contains 84% of the power in the pattern.

For example, if we assume a point source at infinity, the diffraction image is formed at the lens focus $q = f$. If we use IR radiation of $\lambda = 10$ μm, an *F*/2 system forms a blur spot with a diameter of 48 μm. A system operating at *F*/8 will form a blur spot with a diameter of 192 μm. If we use visible radiation of $\lambda = 0.5$ μm, an *F*/2 system forms a spot diameter of 2.4 μm, and a system operating at *F*/8 will form a spot diameter of 9.6 μm.

As seen in Fig. 2.13, we can project the diffraction blur to the object plane in the same way as the detector footprint that was seen in Fig. 2.10. This approach allows us to estimate the diffraction-limited imaging performance in terms of distance to the object plane. The full-angle blur caused by diffraction

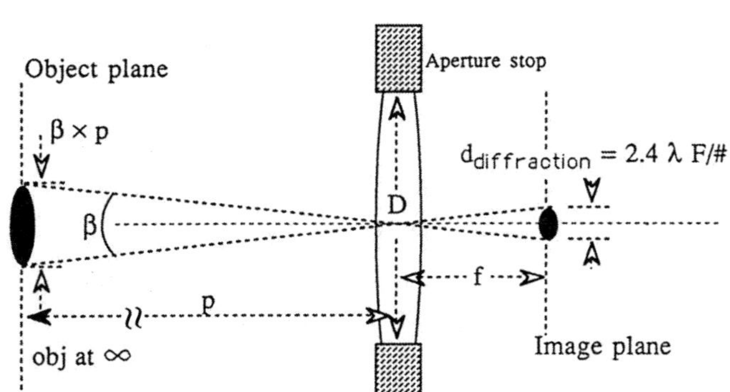

Figure 2.13 Projection of the diffraction-blur spot to the object plane.

is $\beta = 2.4\ \lambda/D$, which is consistent with an image-plane spot diameter of $2.4\ \lambda\ (F/\#)$, as seen in Eq. (2.3):

$$\beta = \frac{2.4\ \lambda\ (F/\#)}{f} = \frac{2.4\ \lambda\ (f/D)}{f} = \frac{2.4\ \lambda}{D}. \tag{2.3}$$

The footprint of the diffraction spot at the object is the diffraction angular blur multiplied by the object distance: $\beta \times p$. A feature of interest in the object whose size is smaller than $\beta \times p$ will not be resolved by the optical system.

In an imaging system that employs a detector array, the best possible object-plane resolution is determined by the convolution of the diffraction-blur footprint with the detector footprint. The width of this convolution will be less than or equal to the addition of the widths of the individual footprints. Any aberrations present in the system will increase the width of the object-plane resolution element, further reducing image quality.

2.7.2 Aberrations

The ideal imaging situation is expressed in Eq. (2.1a), which assumes that all rays originating from a given object point intersect at a common point in the image. In practice, these rays do not all intersect at a single image point, but rather form a finite-sized blur spot, because of the effects of aberrations. The following aberrations are inherent in the performance of systems having spherical surfaces: spherical aberration, coma, astigmatism, distortion, field curvature, and chromatic aberration. These aberrations are inherent in the imaging process itself, arising from the refraction of rays at spherical interfaces, according to Snell's law.

The size of the resulting aberrated blur spot depends on the *F/#* of the imaging lens and on the field angle θ of the object point. Each aberration is characterized by a different functional dependence on *F/#* and field angle. The width of the blur spot caused by aberrations is proportional to the inverse of some power of the *F/#*: $(F/\#)^{-1}$, $(F/\#)^{-2}$, $(F/\#)^{-3}$, etc., and to the field angle raised to some power: θ, θ^2, θ^3, etc.

Because it is not possible to achieve a smaller spot size than what is dictated by diffraction, the size of an aberrated blur spot can be compared to that of the diffraction-blur spot to determine whether diffraction or aberrations dominates the image quality.

2.7.3 Image-quality summary

F-number and FOV affect both flux transfer and aberrations. A small *F/#* (large aperture) and large FOV lead to a high flux-transfer efficiency. But aberrations generally get worse as *F/#* decreases and as FOV increases. Thus, an engineering tradeoff exists between the amount of light reaching the sensors and the quality of the image. If we have extra light reaching the detectors, we can generally trade smaller flux transfer for better image quality. Conversely, if we do not have enough light for good sensor performance, we can sometimes get more light at the expense of a poorer quality image. Low-*F*/# systems have the smallest diffraction-limited spot sizes and thus the best potential performance. However, aberrations generally get worse as *F*/# gets smaller. Thus, low-*F*/# systems are harder to correct to diffraction-limited performance. Because long-λ systems are more affected by diffraction, they are easier to correct to diffraction-limited performance than are short-λ systems.

2.8 Infrared Materials

As seen in Fig. 2.14, the atmosphere has *windows* of high transmittance in the 8- to 14-μm band (long-wave infrared, LWIR), the 3- to 5-μm band (mid-wave infrared, MWIR), the short-wave infrared (SWIR), the near IR (NIR), and the visible band. These transmission windows have variations in transmittance caused by molecular absorption. Photon sensors are generally optimized for operation in only one of these bands because of the relation of bandgap to cutoff wavelength and the relation of bandgap to cooling requirements, which we will cover in Chapter 5. The transmittance of the atmosphere shown in Fig. 2.14 is for an 1800-m horizontal path at sea level, with 17 mm of precipitable water. Transmittance will decrease for longer path lengths, higher humidity, and when aerosols are present.

Figure 2.15 shows the spectral transmittance of several common IR materials for specified material thicknesses. The refractive index n of a

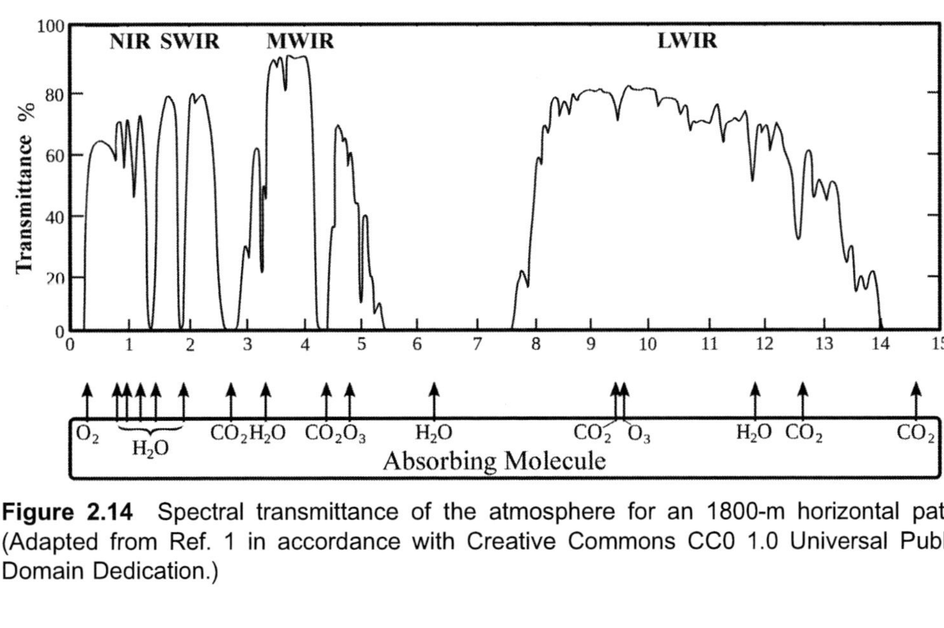

Figure 2.14 Spectral transmittance of the atmosphere for an 1800-m horizontal path. (Adapted from Ref. 1 in accordance with Creative Commons CC0 1.0 Universal Public Domain Dedication.)

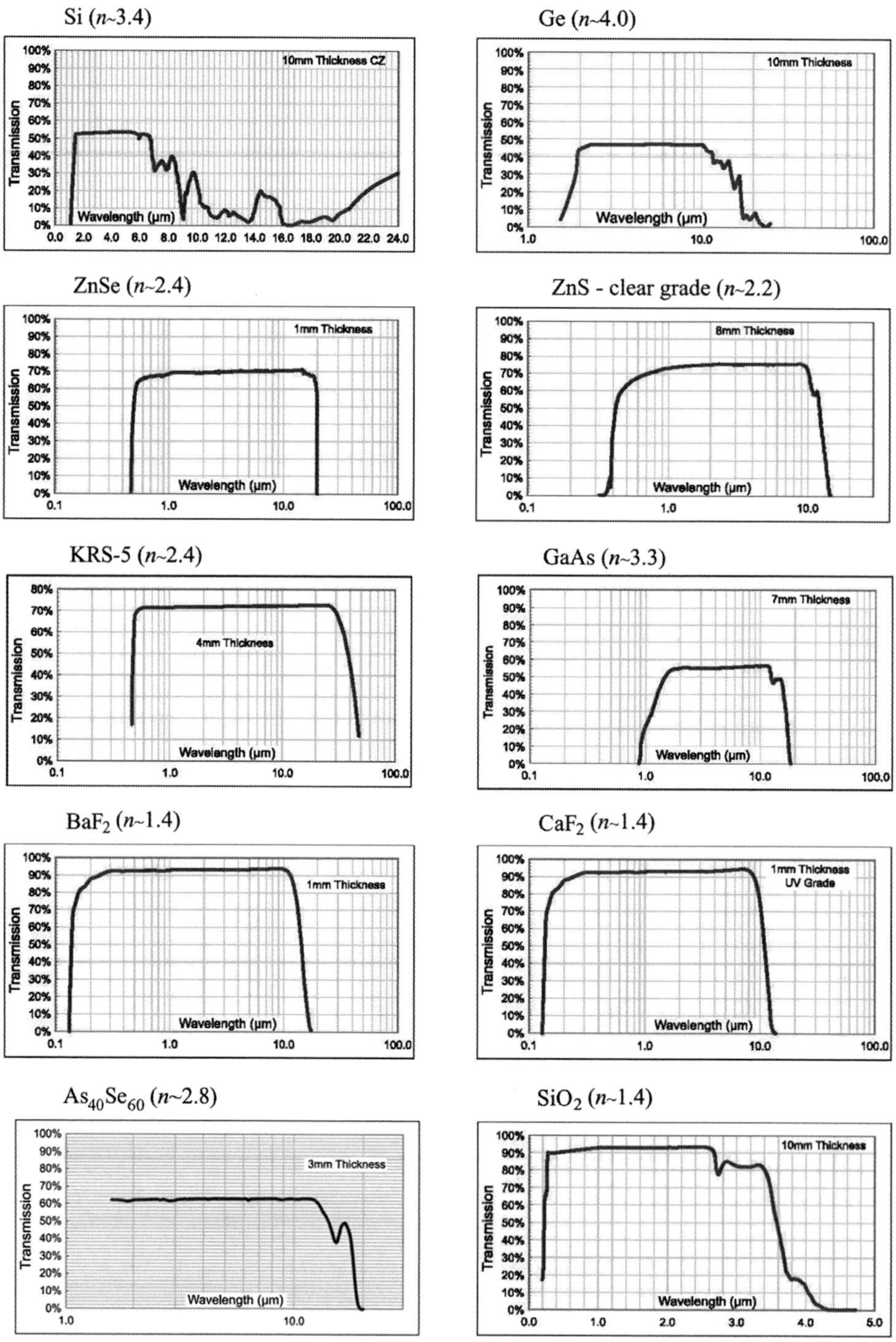

Figure 2.15 Spectral transmittance of various IR materials. (Adapted from Ref. 2.)

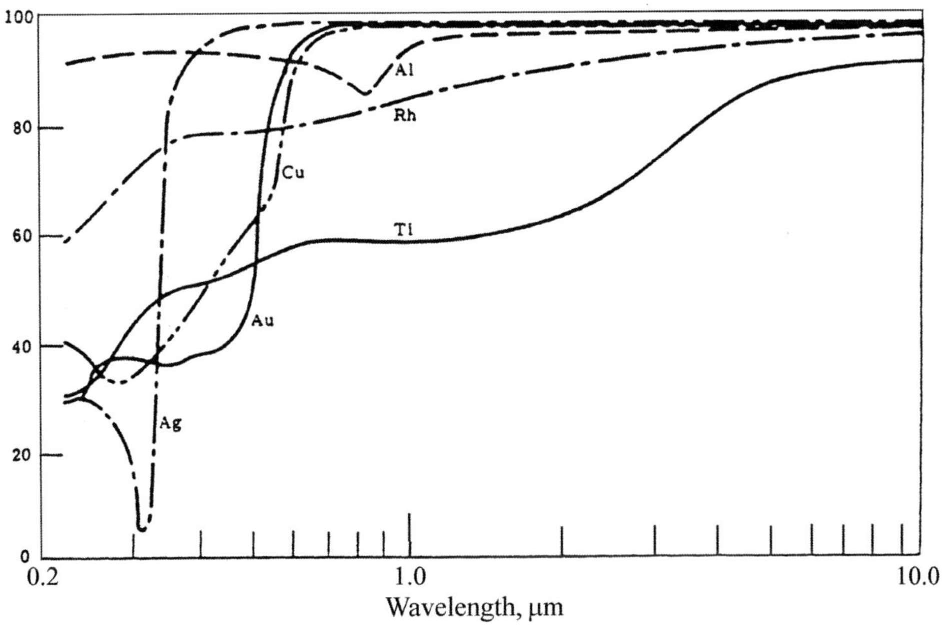

Figure 2.16 Spectral reflectance of various metals: Ag, Al, Au, Cu, Rh, and Ti. (Adapted from Ref. 3.)

material determines its constant transmittance level in the transparent region. This transmittance can be improved using anti-reflection coatings. These coatings are especially important for high-refractive-index IR materials. Molecular absorption determines the fine-scale spectral variations in the transmittance curves, as well as determining the usable transmission band for a given material. In the absorptive spectral regions, the transmittance will have a negative-exponential dependence on material thickness according to Beer's law.

Metals are often used as IR mirrors. In Fig. 2.16, we see the spectral reflectance of several metals across the range from visible to LWIR. In the IR, most metals are good reflectors. In the ultraviolet, all metals absorb, with aluminum being the best reflector. Silver is the best reflector in the visible, but aluminum is also good and is a more environmentally stable material.

2.9 Modulation Transfer Function

The modulation transfer function (MTF) is an alternative method to describe image quality in terms of spatial frequencies, instead of in the spatial coordinates of the impulse response.[4] The smallest image detail that an optical system is capable of forming is the impulse response $h(x)$. The image irradiance (W/cm^2) distribution $g(x)$ is the ideal image $f(x)$ convolved with the impulse response $h(x)$:

$$f(x) * h(x) = g(x). \tag{2.4}$$

The ideal image $f(x)$ is the irradiance distribution in the image plane for a system with perfect image quality, that is, a magnified version of the object scene. Convolution of the ideal image function with the impulse response (with the resulting loss of detail) can be cast in the frequency domain by means of the convolution theorem, which states that a convolution in the spatial domain is equivalent to a product in the frequency domain.

2.9.1 Spatial frequency

The irradiance distribution is composed of *spatial frequencies* in the same way that time-domain electrical signals are composed of various frequencies by means of a Fourier analysis. As seen in Fig. 2.17, by taking a one-dimensional profile across a two-dimensional irradiance distribution, we obtain an irradiance-vs-position waveform that can be Fourier decomposed in the same way as if the waveform was in the more familiar form of volts vs time. The Fourier decomposition is described in terms of spatial frequencies with units of cycles (cy) per unit distance, analogous to temporal frequencies in cy/s for a time-domain waveform. Typically for optical systems, the spatial frequency is in cy/mm. Sometimes we see spatial frequency expressed in line pairs per millimeter (lp/mm). One cycle equals one black–white line pair. An example of a one-dimensional spatial frequency waveform is shown in Fig. 2.18. The spatial period X (crest-to-crest repetition distance) of the waveform can be inverted to find the x-domain spatial frequency denoted by $\xi \equiv 1/X$.

Angular spatial frequency is typically encountered in the specification of imaging systems designed to observe a target at a long distance. In this situation, it is convenient to specify system performance in angular units without needing to specify a particular object distance. Angular spatial frequency is most often specified in cycles/milliradians (cy/mrad). This can initially be a troublesome concept because both cycles and milliradians are

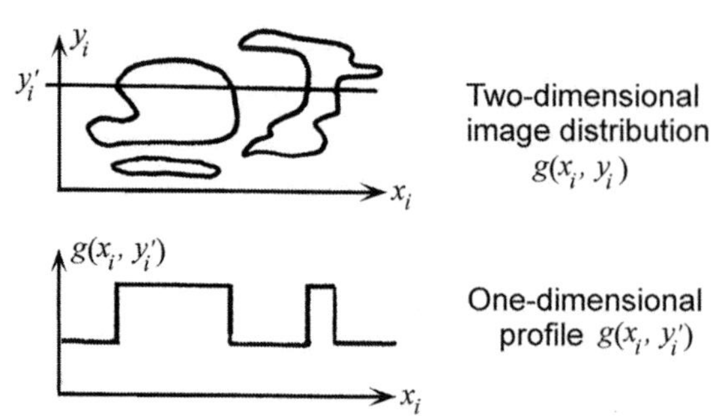

Figure 2.17 Definition of a spatial-domain irradiance waveform.

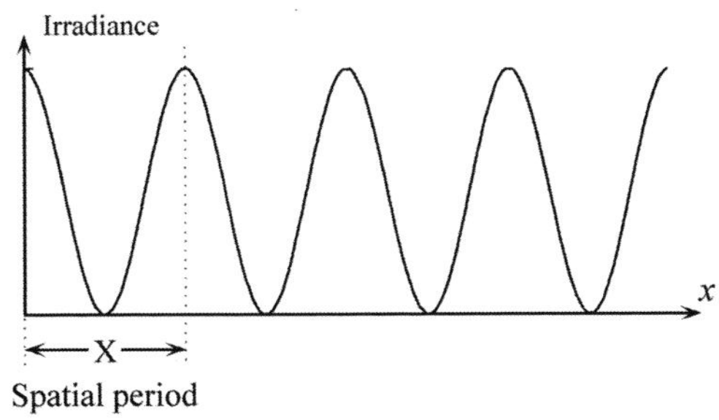

Figure 2.18 One-dimensional spatial frequency.

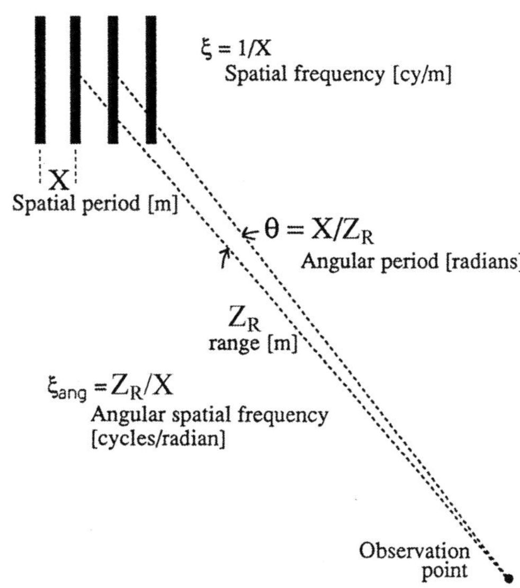

Figure 2.19 Angular spatial frequency.

dimensionless quantities but, with reference to Fig. 2.19, we find that the angular spatial frequency ξ_{ang} is simply the range Z_R multiplied by the target spatial frequency ξ. For a periodic target of spatial period X, we define an angular period $\theta \equiv X/Z_R$, an angle over which the object waveform repeats itself. The angular period is in radians if X and Z_R are in the same units. Inverting this angular period yields angular spatial frequency $\xi_{ang} = Z_R/X$ in cy/rad. Given the typical resolution of optical systems, X is often given in meters and Z_R in kilometers. The ratio Z_R/X is then given in units of cy/mrad.

2.9.2 Modulation transfer function definition

Equation (2.4) describes the loss of detail inherent in the imaging process as the convolution of the ideal image function with the impulse response. As previously noted, a convolution in the spatial domain is equivalent to a multiplication in the frequency domain.[4] Taking the Fourier transform (denoted as $\mathcal{F}$) of both sides of Eq. (2.4) yields Eqs. (2.5) and (2.6):

$$\mathcal{F}\{g(x)\} = \mathcal{F}\{f(x) * h(x)\} \tag{2.5}$$

and

$$G(\xi) = F(\xi) \times H(\xi). \tag{2.6}$$

The uppercase functions denote the Fourier transforms of the corresponding lowercase functions: F denotes the object spectrum; G denotes the image spectrum; and H denotes the spectrum of the impulse response. The magnitude of $H(\xi)$ is the modulation transfer function; it relates the object and image spectra multiplicatively. The Fourier transform changes the irradiance waveform from a function of spatial position to a function of spatial frequency but does not generate any new information. The appeal of the frequency-domain viewpoint is that the multiplication of Eq. (2.6) is easier to perform and visualize than the convolution of Eq. (2.5). This convenience is most apparent in the analysis of imaging systems consisting of several subsystems. As Eq. (2.7) demonstrates, each subsystem has its own transfer function, which is the Fourier transform of its impulse response.

The result of all the subsystems operating on the input object distribution is a multiplication of their respective transfer functions. Figure 2.20 illustrates that we can analyze a combination of several subsystems by the multiplication of transfer functions of Eq. (2.8) rather than the convolution of impulse responses of Eq. (2.7):

$$f(x) * h_1(x) * h_2(x) * \ldots * h_n(x) = g(x) \tag{2.7}$$

and

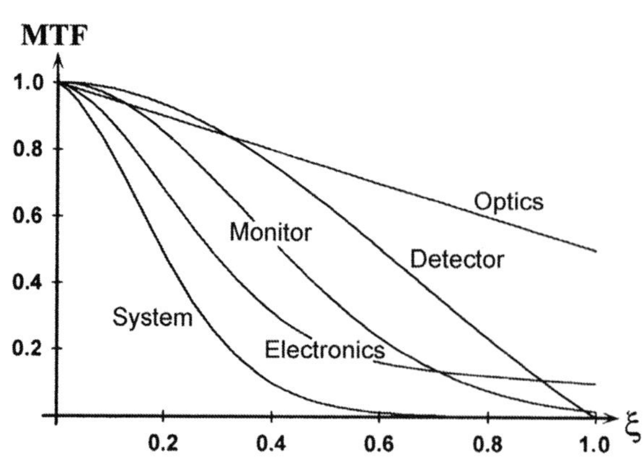

Figure 2.20 The aggregate transfer function of several subsystems is a multiplication of their transfer functions.

Figure 2.21 Sinewave target containing several spatial frequencies.

$$F(\xi) \times H_1(\xi) \times H_2(\xi) \times \ldots \times H_n(\xi) = G(\xi). \quad (2.8)$$

The modulation transfer function is the magnitude response of the optical system to sinusoids of different spatial frequencies. When we analyze an optical system in the frequency domain, we consider the imaging of sinewave inputs (Fig. 2.21) rather than point objects. Sinewave targets are typically printed as diffuse reflective targets intended to be viewed under conditions of natural illumination. Alternatively, sinewaves of irradiance can be generated by diffraction as Young's fringes or by using two-beam interference techniques.

A linear shift-invariant optical system images a sinusoid as another sinusoid. The finite spatial extent of the optical system's impulse response causes a filling in of the valleys and a lowering of the peak levels of the sinusoid. The limited spatial resolution of the optical system results in a decrease in the modulation depth $\boldsymbol{m}$ of the image waveform relative to what was in the object waveform (Fig. 2.22). Modulation depth is defined as the amplitude of the sinusoidal irradiance variation divided by the irradiance bias level:

$$\boldsymbol{m} = \frac{E_{\max} - E_{\min}}{E_{\max} + E_{\min}} = \frac{2 \times \text{ac component}}{2 \times \text{dc component}} = \frac{\text{ac}}{\text{dc}}. \quad (2.9)$$

We can see from Fig. 2.22 that modulation depth is a measure of contrast, with $\boldsymbol{m} \to 0$ as $A_{\max} - A_{\min} \to 0$ and $\boldsymbol{m} \to 1$ as $A_{\min} \to 0$. The $\boldsymbol{m} = 0$ condition means that no spatial variation of irradiance exists in the image. However, the $\boldsymbol{m} = 0$ condition does not imply that the image-irradiance level is zero. Another consequence of Eq. (2.9) is that the $\boldsymbol{m} = 1$ unit-modulation-depth condition is obtained when the irradiance waveform has a minimum value of zero, regardless of the maximum irradiance level. This is a consequence of the usual non-radiometric normalization of the MTF. In practice, the variations of irradiance must be sufficient to be seen above any noise level present. The interpretation of modulation depth as a measure of contrast is that a waveform with a small modulation depth, as seen in

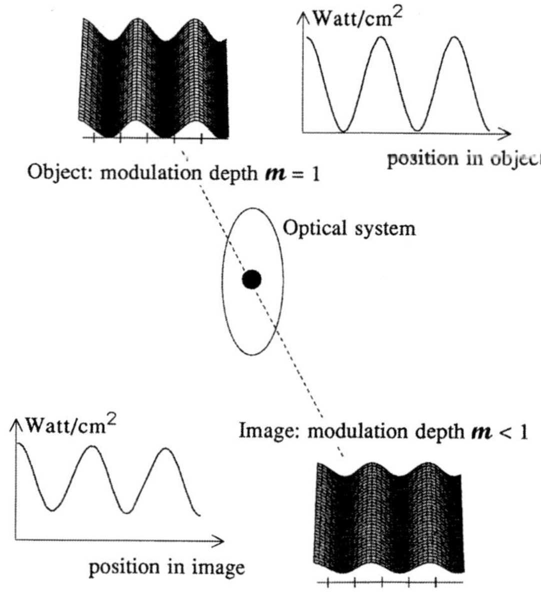

Figure 2.22 Modulation depth decreases going from object to image.

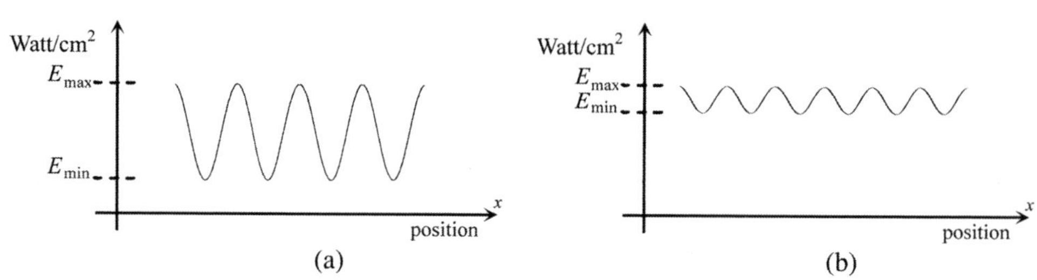

Figure 2.23 Definition of modulation depth per Eq. (2.9): (a) high contrast and (b) low contrast.

Fig. 2.23, is more difficult to distinguish against a uniform background irradiance than is a waveform with a larger modulation depth, especially with some level of noise present.

The MTF is spatial-frequency dependent. The limited resolution of the optics is more important at high spatial frequencies, where the scale of the detail is smaller. The MTF is generally a decreasing function of spatial frequency, as seen in Fig. 2.24. In the situation shown, where the object modulation depth is constant, the MTF is simply the image modulation as a function of spatial frequency:

$$\mathrm{MTF}(\xi) \equiv m_{\mathrm{image}}(\xi)/m_{\mathrm{object}}. \tag{2.10}$$

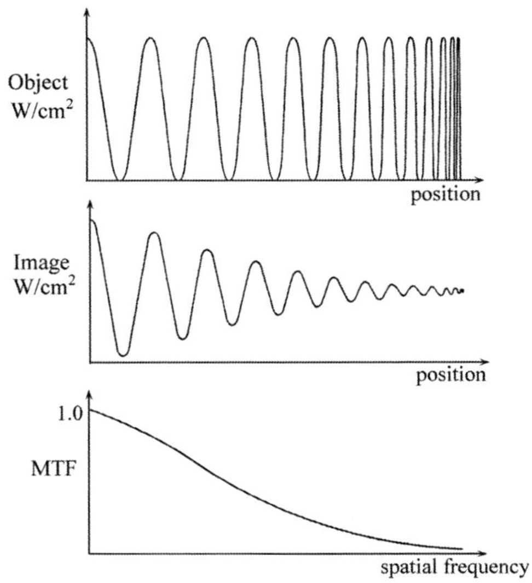

Figure 2.24 Modulation transfer function is the decrease of modulation depth with increasing frequency.

2.9.3 MTF examples

We will need two specific MTFs in our discussion of IR imaging systems. Diffraction-limited MTFs for systems with circular or square aperture are seen in Fig. 2.25. These plots provide an easy-to-calculate upper limit to performance, since the calculation requires only λ and *F*/#. The same conditions that create a smaller diffraction spot (shorter wavelength and smaller *F*/#) will increase the cutoff frequency ξ_{cutoff}. A system cannot perform better than the diffraction-limited MTF, and any aberrations present in the imaging system will pull the MTF curve downward.

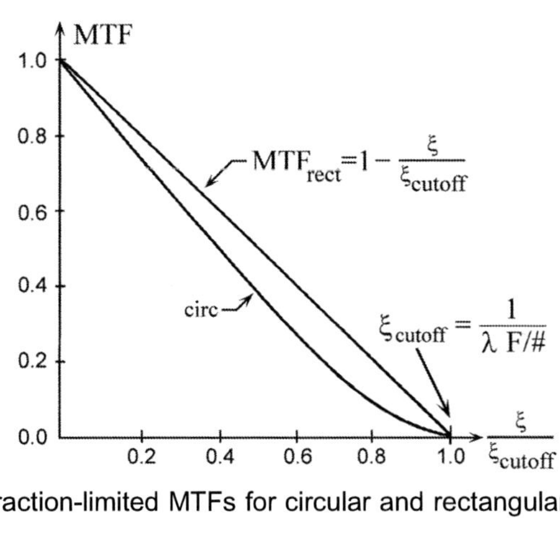

Figure 2.25 Diffraction-limited MTFs for circular and rectangular aperture systems.

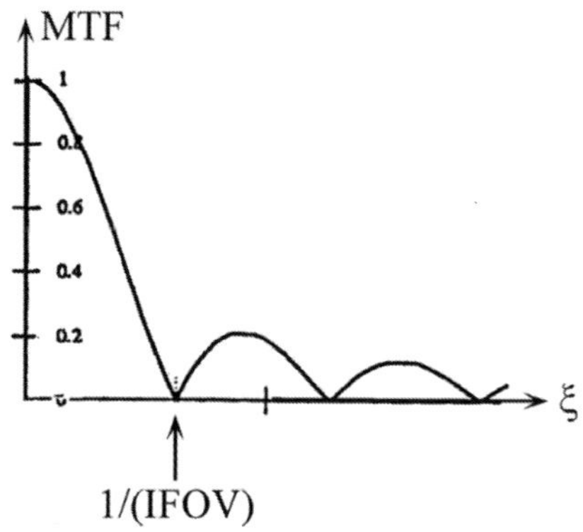

Figure 2.26 Detector footprint MTF.

The detector-footprint MTF is a sinc function,[5] with its first zero at $\xi = 1/\mathrm{IFOV}$ (as seen in Fig. 2.26):

$$\mathrm{MTF}_{\mathrm{footprint}}(\xi) = \left| \frac{\sin(\pi\,\xi\ \mathrm{IFOV})}{(\pi\,\xi\ \mathrm{IFOV})} \right|. \tag{2.11}$$

References

1. https://commons.wikimedia.org/wiki/File:Atmosfaerisk_spredning-ru.svg
2. A. Symmons and M. Lifshotz, *Field Guide to Infrared Optical Materials*, SPIE Press, Bellingham, Washington (2021) [doi: 10.1117/3.2589608].
3. W. Wolfe and G. Zissis, *The Infrared Handbook, Revised Edition*, SPIE Press, Bellingham, Washington (1985).
4. G. Boreman, *Modulation Transfer Function in Optical & Electro-Optical Systems, 2nd Edition*, SPIE Press, Bellingham, Washington (2021) [doi: 10.1117/3.2581373].
5. J. Gaskill, *Linear Systems, Fourier Transforms, and Optics*, Wiley, New York (1978).

Bibliography

Fischer, R., B. Tadic-Galeb, and P. Yoder, *Optical System Design, 2nd Edition*, McGraw Hill, New York (2008).

Hecht, E., *Optics, 5th Edition*, Pearson, London (2015).

Jenkins, F., and H. White, *Fundamentals of Optics, 4th Edition*, McGraw Hill, New York (2001).

Smith, W., *Modern Optical Engineering, 4th Edition*, McGraw Hill, New York (2007).

Chapter 3
Radiometry

3.1 Introduction

Radiometry is a framework to quantitatively describe flux transfer through an optical system. The fundamental question to be answered, given the configuration of the optical system and the source, is how much power radiated by the source is collected at the detector surface? Radiometric calculations allow the prediction of a system's signal-to-noise ratio (SNR); the radiometric efficiency of the system is crucial to the overall SNR and image detectability. We consider here the radiometry of incoherent thermal sources, ignoring any interference effects. The power distribution over any surface is calculated as a scalar sum of incremental powers. Diffraction is ignored, except for the special case of unresolved sources, and we assume that light travels in straight lines. Small angles ($\sin\theta \approx \theta$) are assumed. Thus, we consider paraxial radiometry in the same spirit as paraxial optics: a first-order solution that is easy to calculate.

3.1.1 Solid angle

Solid angle, denoted by Ω, measures a range of directions in three-dimensional space going from a point location to a surface located some distance away. Solid angle is defined as the ratio A/r^2, where A is the area on the surface of a sphere of radius r, that is defined by the cone of rays that include the solid angle Ω from the center of the sphere. Solid angle is, strictly speaking, a dimensionless quantity. For convenience, we can associate the "units" of steradians (sr) with this A/r^2 quantity. To understand solid angle, it is useful to recall the definition of radians from plane geometry, as seen in Fig. 3.1. The angle θ in radians is the dimensionless ratio s/r, where s is the arc length on the circumference of the circle that is defined by the two rays that include the angle θ, and r is the radius of the circle. The planar angle θ in radians measures the range of in-plane pointing directions from the center point of the circle to the arc s.

Under this defintion, a full sphere, having $4\pi r^2$ of surface area, subtends 4π sr of solid angle from the center of the sphere, and a hemisphere subtends

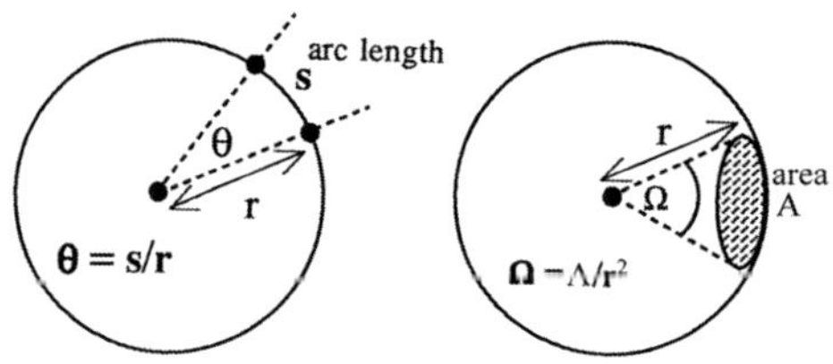

Figure 3.1 Definitions of (left) planar angle and (right) solid angle.

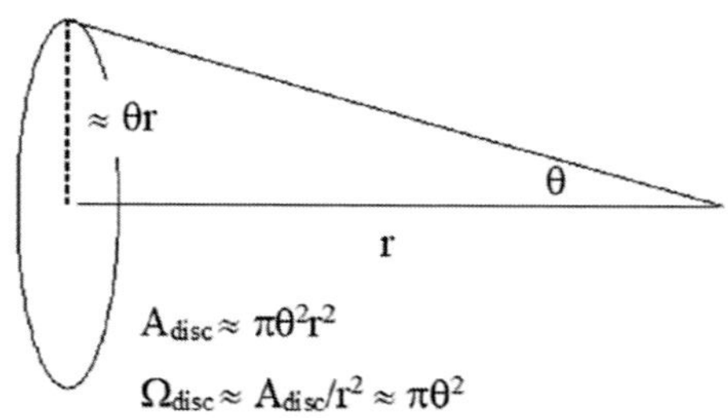

Figure 3.2 The relation between planar half angle θ and solid angle Ω.

$\Omega = 2\pi$ sr. Although solid angle is defined in terms of area on the surface of a sphere, within the small-angle approximation, we can write

$$\Omega = A/r^2 \tag{3.1}$$

even for areas A that are flat rather than a portion of a sphere, if the area in question is small compared to r^2. Another useful small-angle approximation relates planar half-angle subtense θ to solid angle Ω as $\Omega \approx \pi\theta^2$, as seen in Fig. 3.2.

A small flat disc is a good representation for the entrance aperture of an optical system, and a small flat square can model a detector. Also useful is an expression for the solid angle of a small area A that is tilted with respect to the line of sight (the line from the starting point to the center of the surface), seen in Fig. 3.3. If the area is tilted at an angle θ (measured from the surface normal to the line of sight), its solid-angle subtense Ω is reduced by a factor of cosθ from what it would be if it were not tilted, according to Eq. (3.2). The term $A\cos\theta$ is sometimes referred to as the *projected area*.

$$\Omega = (A\cos\theta)/r^2. \tag{3.2}$$

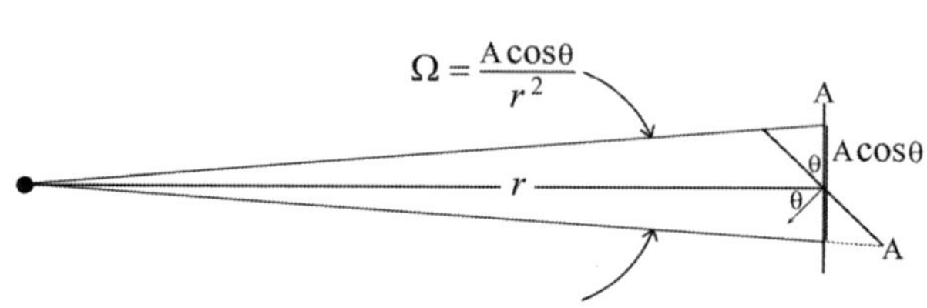

Figure 3.3 Projected area and solid angle of a tilted surface.

3.1.2 Radiometric nomenclature and units

We use a set of nomenclature defined in metric units as seen in Table 3.1. Understanding the units is key to making correct radiometric calculations. We will consider calculational examples of each quantity.

The units in Table 3.1 are called *energy derived* because they are based on the amount of energy. An analogous set of quantities is based on the number of photons. These *photon-derived* quantities are written with subscript q. For instance, ϕ_q is photon flux and is measured in photon/s. Similarly, I_q is photon intensity, in units of photon/(s sr). A conversion between the two sets of units is easily done using Eq. (3.3), the formula for energy carried per photon:

$$\mathcal{E} = hc/\lambda \text{ [J/photon]}, \tag{3.3}$$

where Planck's constant $h \approx 6.6 \times 10^{-34}$ J s, and the speed of light $c \approx 3 \times 10^8$ m/s. For example, the conversion for flux is ϕ [J/s] = ϕ_q [photon/s] × $\mathcal{E}$ [J/photon]. Photon-derived units are useful when considering a *photon detector*, which responds directly to photon events, rather than a sensor that responds to the thermal energy of the photons (a *thermal detector*). Note that $\mathcal{E}$, and hence the conversion between the two sets of units, depends on λ because a long-wavelength photon carries less energy than does a short-wavelength photon. A photon with $\lambda = 10$ μm carries $\approx 2 \times 10^{-20}$ J/photon, while a photon with $\lambda = 0.5$ μm carries $\approx 4 \times 10^{-19}$ J/photon. Because $\mathcal{E}$ is interpreted as [J/photon] ≡ [W/(photon/s)], we can also interpret $1/\mathcal{E} = \lambda/(hc)$ as the number of photons/s it takes to make one watt. For a wavelength $\lambda = 10$ μm, $\lambda/(hc) = 5 \times 10^{19}$ (photons/s)/W, while at $\lambda = 0.5$ μm, $\lambda/(hc) = 2.5 \times 10^{18}$ (photon/s)/W. It takes more photons/s to make a watt at longer wavelengths.

3.2 Irradiance and Exitance

Irradiance E and exitance M both have units of W/cm^2, which is a spatial power density. However, these two quantities have a different physical interpretation. Irradiance is the power per unit area falling on a surface,

Table 3.1 Radiometric quantities.

Symbol	Quantity	Units
Q	Energy	J
ϕ	Flux (power)	W = J/s
I	Intensity	W/sr
E	Irradiance	W/cm^2
M	Exitance	W/cm^2
L	Radiance	W/(cm^2 sr)

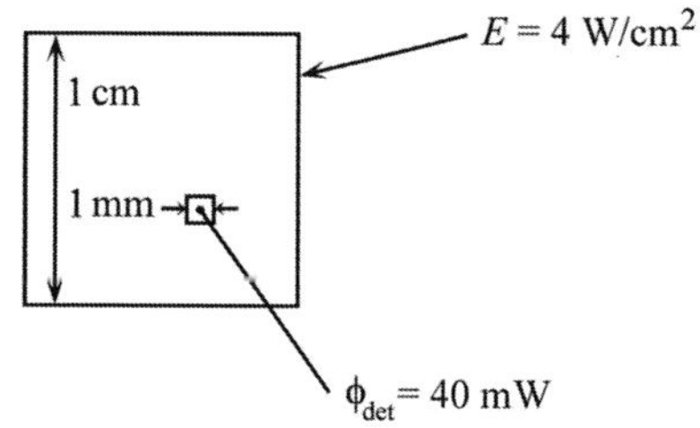

Figure 3.4 Calculation example for irradiance.

characterizing a receiver or other passive surface. Exitance is the power per unit area leaving a surface, used to describe a self-luminous surface or a diffusely reflective surface acting as a source. As an initial example, consider an irradiance, $E = 4$ W/cm^2, uniformly distributed over a 1-cm^2 surface, as seen in Fig. 3.4. How much power falls on a detector which is a 1-mm^2 portion of the surface? If 4 W fall onto each cm^2 of the surface, then 40 mW are received on a 1-mm^2 area, consistent with

$$\phi_{\mathrm{d}}\,[\mathrm{W}] = E\,[\mathrm{W/cm^2}] \times A_d\,[\mathrm{cm^2}]. \tag{3.4}$$

The simple expression of Eq. (3.4) is valid as long as the irradiance is uniform across the detector surface. If the irradiance is nonuniform across the receiving surface, the irradiance must be integrated over the surface to obtain the power received. In radiometric calculations for imaging systems, the result is often cast in terms of image-plane irradiance. This formulation is convenient because from Eq. (3.4) we multiply this irradiance by the sensor area to calculate the power received on the detector.

Alternatively, exitance is used to calculate the total power radiated from an emitting surface. If a radiating surface of area 1 mm^2 has an exitance of $M = 3$ W/cm^2, what is the total radiated power? Because 3 W radiate from each cm^2 of the source, 30 mW are radiated (into a hemisphere) from a 1-mm^2 source area, according to

$$\phi_{\mathrm{total}}\,[\mathrm{W}] = M\,[\mathrm{W/cm^2}] \times A_{\mathrm{source}}\,[\mathrm{cm^2}]. \tag{3.5}$$

In Eq. (3.5), we consider that A_{source} is the *forward* area; we assume that the source radiates from one side only (the side toward the optical system or the receiver). The power calculated is that which would be the total collected by the hemisphere that encloses the forward direction from the source, as seen in Fig. 3.5. As such, exitance is inherently an angle-averaged quantity. Exitance does not allow calculation of irradiance as a function of angle on the hemisphere (for a flat source, the hemisphere is not uniformly illuminated). To obtain angle-resolved information about the source, we must use the quantity radiance.

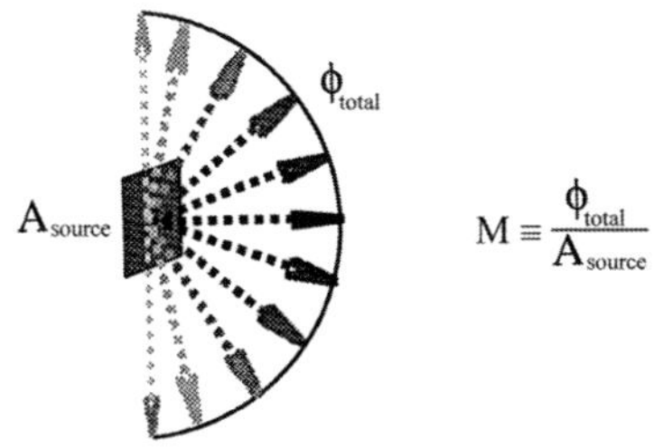

Figure 3.5 Calculation example for exitance.

3.3 Intensity

The quantity intensity I has units of W/sr and is interpreted as the power radiated from a source per unit solid angle. To characterize a point source, the quantity intensity must be used because intensity does not involve units of area. Intensity may also be used to characterize extended sources. As a reminder (Section 2.7), a point source is not resolved by the optical system. The dimensions of the source are smaller than the projection of the resolution element to the object plane.

The first calculation with intensity is to determine how many watts are radiated by a point source with a given intensity, such as 3 W/sr. The total solid angle into which a point source radiates is 4π sr (a sphere) so, according to

$$\phi\,[\mathrm{W}] = I\,[\mathrm{W/sr}] \times \Omega\,[\mathrm{sr}], \tag{3.6}$$

the total radiated power in this case is 12π W. However, in practice, we typically do not enclose the point source within a sphere. More often, the quantity of concern is how much power is radiated through an aperture of a certain size placed at a certain distance from the source, as illustrated in Fig. 3.6.

How much power will the point source with $I = 3$ W/sr radiate through the circular aperture in an opaque screen? We assume that the aperture has a radius of 1 mm, and we place the screen 3 cm from the point source. We calculate the solid angle subtended by the aperture as seen from the point source, according to Eq. (3.1): $\Omega_{aperture} = A_{aperture}/r^2 = \pi\,(10^{-3}\,\mathrm{m})^2/(3 \times 10^{-2}\,\mathrm{m})^2 = 3.5 \times 10^{-3}$ sr. Then we multiply this solid angle by the intensity of the source to obtain the power radiated through the aperture as $\phi = I \times \Omega_{aperture} = 10.5$ mW.

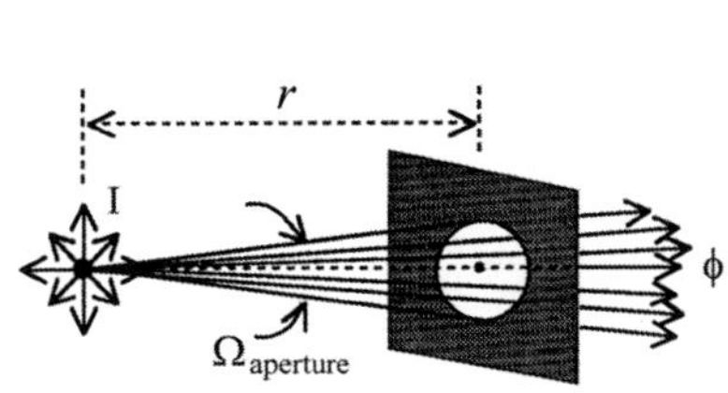

Figure 3.6 Calculation example for intensity.

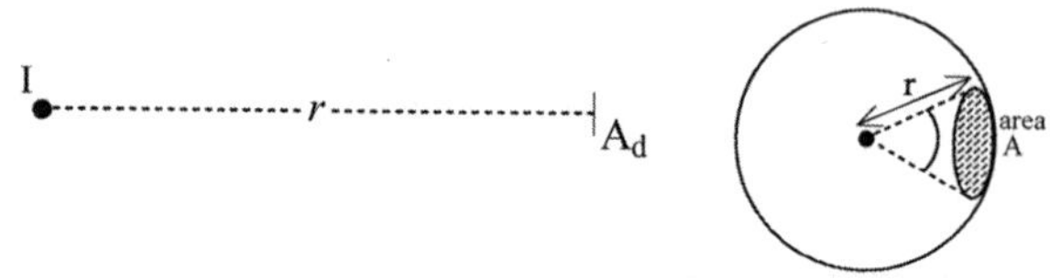

Figure 3.7 Inverse-square law irradiance falloff for a point source.

3.3.1 Inverse-square law for point sources

Given an amount of power ϕ radiated from a point source, that same power passes through any sphere centered on the source. Thus, the irradiance $E = \phi/A = \phi/(4\pi r^2)$ falls as $1/r^2$, as seen in Fig. 3.7. In this calculation, the units can be confusing. We begin with the Eq. (3.6): $\phi = I \times \Omega$, where no ambiguity exists with the units because I is in W/sr, and Ω is in sr. However, when A_d/r^2 is substituted for Ω, A_d and r^2 must be in the same units to yield (dimensionless) sr. Confusion can arise when $\phi = I\ A_d/r^2$ is divided by A_d, which yields irradiance (W/area) on the left-hand side of the expression. The sr on the right-hand side can be considered to cancel if the units of area in the irradiance and the units of r^2 are the same. So, if r is measured in meters, then E is in W/m^2; if r is measured in centimeters, then E is in W/cm^2.

3.3.2 Irradiance on a tilted receiver

As seen in Fig. 3.8, to find the irradiance from a point source on a small flat surface tilted with respect to the line of sight by an angle θ_d, we first evaluate how much power falls on the surface and then use $E \approx \phi/A_d$. Using the expression for the solid angle of a tilted receiver of Eq. (3.2), we calculate the power received by the tilted detector as $\phi_d = I\,\Omega_d = I\,A_d \cos\theta_d/r^2$. Then, using Eq. (3.4), we express the irradiance as $E = \phi_d/A_d = I \cos\theta_d/r^2$. The flux captured by the detector becomes smaller as it is tilted, while the detector area does not change, so the irradiance drops.

To calculate irradiance as a function of position on an extended receiver (such as a wall), we use a similar method, illustrated in Fig. 3.9. By letting the receiver area be small, we can approximately calculate irradiance as a *point quantity*, that is, a continuous function of position. However, the orientation of the receiver surface matters. We divide the wall into incremental areas and calculate the power transferred to each one using Eq. (3.2) for the solid angle of a tilted receiver, where r is understood to be the line-of-centers distance. We then divide the power collected on the incremental receiver by the area of that receiver to obtain an expression for irradiance at the center of that

Figure 3.8 Calculation of irradiance on a tilted receiver.

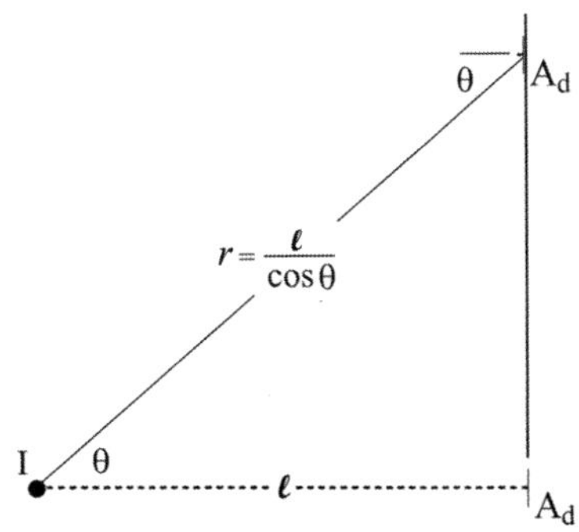

Figure 3.9 Calculation of irradiance from a point source as a function of position on a large flat receiver.

incremental receiver. This calculation technique is first illustrated using a point source and a later example will treat the case of a small flat source.

Consider the case of a point source and a large flat receiver, as shown in Fig. 3.9. Let the vertical direction be taken as the x axis. Take ℓ as the minimum distance between the point source and the plane containing the receiver. The line-of-centers distance r between the point source and the incremental area A_d is $\ell/\cos\theta$. This distance comes into the calculation of the solid angle of the receiver, as seen from the point source.

We calculate irradiance on the wall as a function of the angle θ as follows. For irradiance at the perpendicular distance, we find $E(x=0)=\phi/A_d=I\Omega_d/A_d=\{IA_d/\ell^2\}/A_d=I/\ell^2$. As θ increases, the irradiance will drop via $E(x\neq 0)=\phi/A_d=I\Omega_d/A_d=\{IA_d\cos\theta/(\ell/\cos\theta)^2\}/A_d$. The final dependence of irradiance on θ is thus $E(x\neq 0)=(I/\ell^2)\cos^3\theta=E(x=0)\cos^3\theta$.

3.4 Radiance

Radiance L, with units of W/(cm^2 sr), is used to characterize extended sources. Radiance is analogous to the perceived brightness of the source. A Lambertian source, as seen on the left side of Fig. 3.10, has a brightness

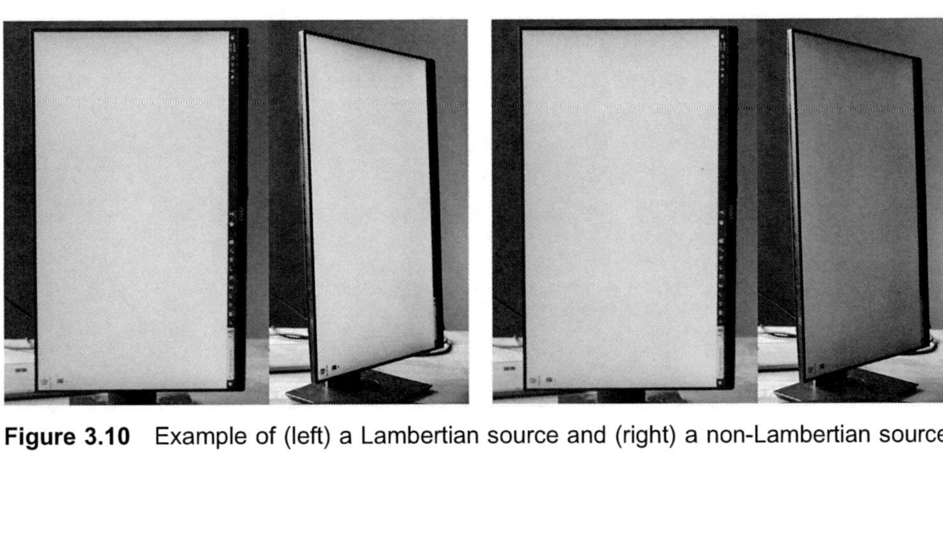

Figure 3.10 Example of (left) a Lambertian source and (right) a non-Lambertian source.

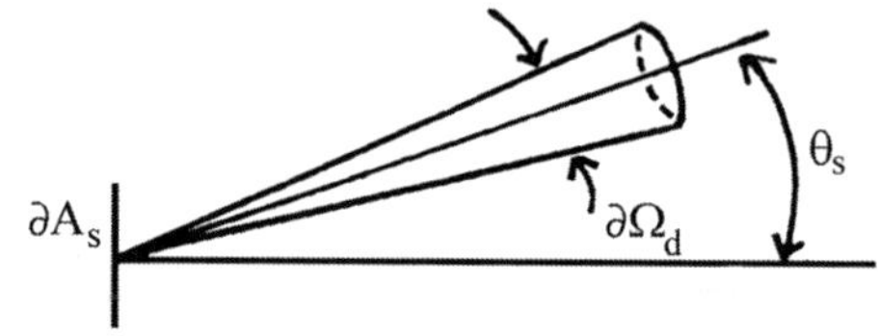

Figure 3.11 Definition of radiance.

that does not vary as the source is tilted. The non-Lambertian source on the right has the characteristic that its brightness changes as the source is tilted.

As shown in Fig. 3.11, radiance is defined for a particular ray direction as the radiant power per unit *projected* source area (perpendicular to the ray), per unit solid angle:

$$L = \frac{\partial^2 \phi}{\partial A_s \cos\theta_s \partial\Omega_d}. \tag{3.7}$$

The term $\partial^2\phi$ is the power radiated into a cone that defines the solid angle $\partial\Omega_d$ of a receiver, which is incremental with respect to both the projected area of the source $\partial A_s \cos\theta_s$ and the solid angle of the receiver $\partial\Omega_d$. The power in the cone $\partial^2\phi$ has a cosine dependence on θ_s, even if L is independent of θ (a Lambertian radiator) because radiance is defined in terms of projected source area. Cross multiplying Eq. (3.7), we find that

$$\partial^2\phi = L\,\partial A_s \cos\theta_s \partial\Omega_d. \tag{3.8}$$

For a small but finite source area A_s and a small but finite solid angle of the detector Ω_d, we can approximate Eq. (3.8) as

$$\phi = L\,A_s \cos\theta_s\,\Omega_d. \tag{3.9}$$

For the simplest calculational example of radiance, let the view angle $\theta_s = 0$, as seen in Fig. 3.12. Assume an extended source of area 1 cm^2, specified by a radiance of 5 W/(cm^2 sr). Located at a distance of 10 m is a 1-mm^2 detector. How much power falls on the detector? We first calculate the solid angle of the detector as $\Omega_d = A_d/r^2 = (1 \times 10^{-3}\text{ m})^2/(10\text{ m})^2 = 10^{-8}$ sr. Then, we multiply this solid angle by the area of the source and the radiance of the source to obtain the power on the detector,

$$\phi_{det} = L \times A_s \times \Omega_d, \tag{3.10}$$

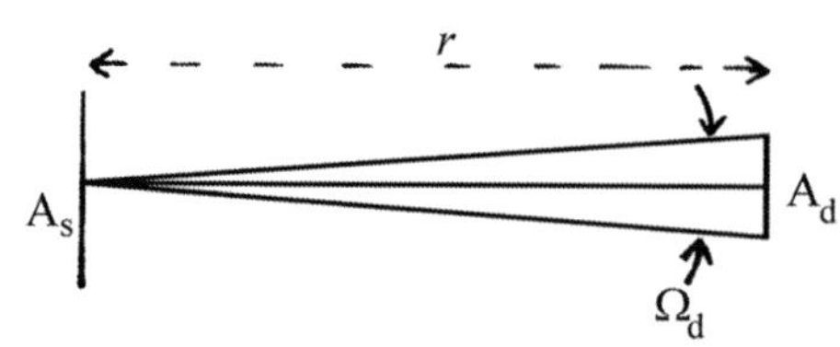

Figure 3.12 Calculation with radiance.

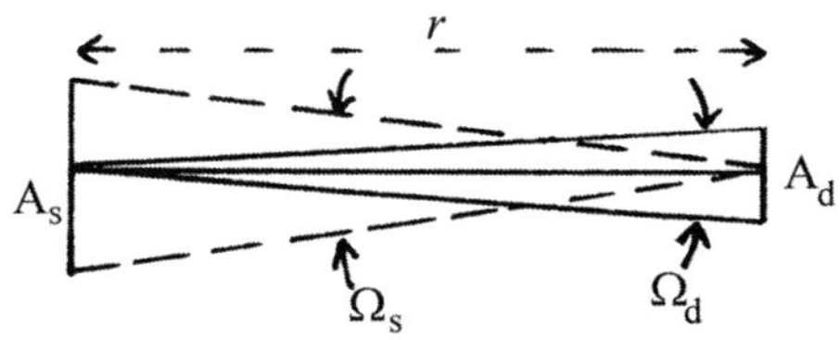

Figure 3.13 Two equivalent $A\Omega$ products.

which yields $\phi_{det} = 5$ W/(cm^2 sr) $\times$ (1 cm)$^2 \times 10^{-8}$ sr $= 5 \times 10^{-8}$ W. Rewriting Eq. (3.10) under the assumption of small angles, we can group the r^2 with either the source or detector area and thus obtain two equivalent expressions in terms of area solid-angle ($A\Omega$) product, $A_s \times \Omega_d = A_d \times \Omega_s$, shown in Eq. (3.11) and Fig. 3.13:

$$\phi_{det} = L \times A_s \times \Omega_d = \frac{L\,A_s\,A_d}{r^2} = L \times \Omega_s \times A_d. \tag{3.11}$$

Either $A\Omega$ product can be used, which often provides a convenient flexibility in calculations. Sometimes one $A\Omega$ product is more conveniently calculated than the other, as illustrated in Fig. 3.14.

If we know the source angular subtense Ω_s, and we want to calculate irradiance at the detector, we can begin with $\phi = L\ A_d\ \Omega_s$. Dividing through by A_d directly gives the desired result of $E = \phi/A_d = L\ \Omega_s$. The irradiance at the plane of the detector depends on the brightness of the source and the solid angle of the source as seen from the observation plane. The irradiance is not directly dependent on source-to-detector distance. A source of a given brightness and of the same angular subtense will produce the same irradiance, regardless of the distance from the source to the detector.

The $A\Omega$ product can be thought of as counting up the number of ways that rays can get through the system, as illustrated in Fig. 3.15. Conceptually,

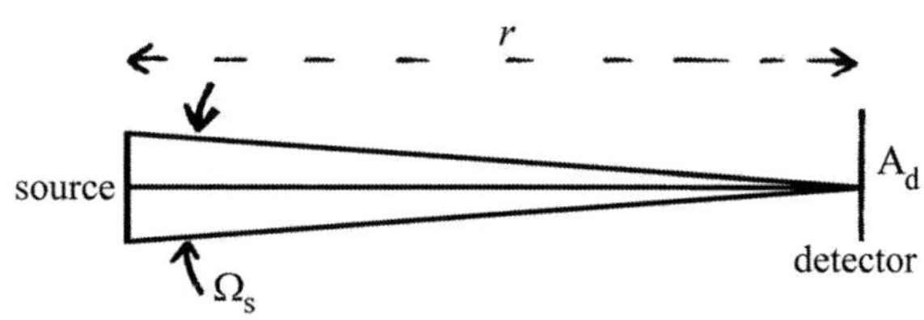

Figure 3.14 Using solid angle of the source to calculate irradiance at the detector.

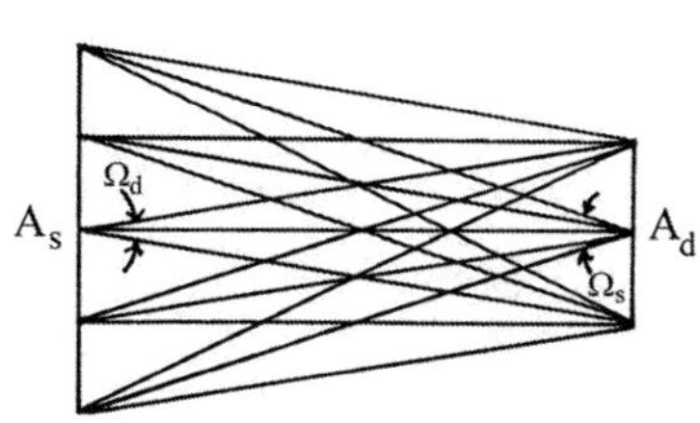

Figure 3.15 Interpretation of the $A\Omega$ product as throughput.

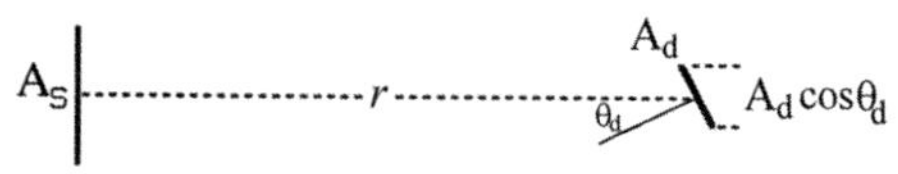

Figure 3.16 Radiance example for a tilted receiver.

we associate a certain amount of power with each ray so that the $A\Omega$ product is interpreted as a flux-gathering capability. Consistent with Eq. (3.11), the source radiance multiplied by the $A\Omega$ product gives the power transferred from the source to the receiver. A bigger $A\Omega$ product means more power is transferred from the source to the receiver for a given source radiance. In this context, the $A\Omega$ product is also called *throughput*.

3.4.1 Radiance example for a tilted receiver

Using Eq. (3.2) for the solid angle of a tilted receiver, we can develop an expression for the case of flux transfer to an on-axis tilted detector, as seen in Fig. 3.16. In this case, the view angle to the source $\theta_s = 0$ and the view angle to the detector $\theta_d \neq 0$.

Beginning with Eq. (3.10), the power transferred to the detector has decreased by the factor of $\cos\theta_d$ compared to the result of Eq. (3.11) so that $\phi_{det} = L \times A_s \times \Omega_d = L \times A_s \times A_d \cos\theta_d / r^2$. Similar to the situation in Fig. 3.8, the irradiance on the detector is reduced by the same factor, yielding $E_d = \phi_{det}/A_d = L \times A_s \times \cos\theta_d / r^2$.

3.4.2 Lambertian radiator

We will now consider situations where the view angle to the source $\theta_s \neq 0$. We make the assumption of a Lambertian radiator, with the source radiance independent of θ_s. Recall the expression Eq. (3.9) for power radiated into a small solid angle Ω_d, from a small extended source of area A_s: $\phi = L A_s \cos\theta_s \Omega_d$. As seen in Fig. 3.17, increasing θ_s decreases $\cos\theta_s$, therefore reducing the projected source area. For L independent of θ_s, the power ϕ radiated into the (constant-solid-angle) cone decreases directly in proportion to the decreasing projected area of the source. For a flat Lambertian source, the intensity $I \equiv \phi/\Omega_d \equiv L\ A_s \cos\theta_s$. Thus, the intensity falls in proportion to $\cos\theta_s$. The intensity of a non-Lambertian source generally has a faster-than-cosine falloff with θ_s.

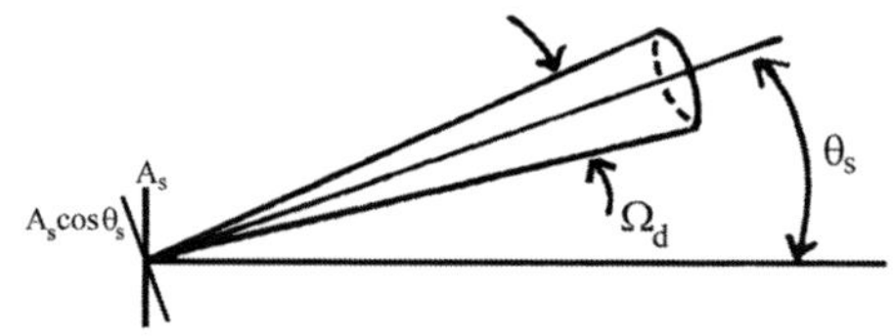

Figure 3.17 Intensity falls as $\cos\theta_s$ for a flat Lambertian source.

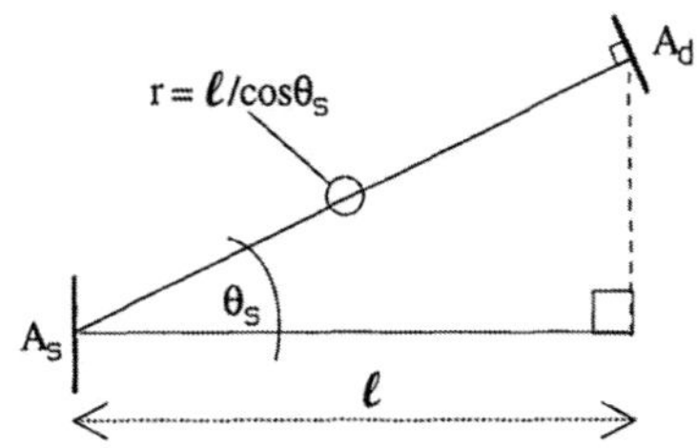

Figure 3.18 Example of flux transfer for a Lambertian source with $\theta_s \neq 0$.

3.4.3 Nonzero view angles

We now consider flux transfer from a small flat Lambertian source to a receiver at a nonzero view angle with respect to the source. Considering the case where $\theta_s \neq 0$ and $\theta_d = 0$ (the detector's normal is along the line of centers), we find a $\cos^3\theta_s$ dependence of flux received by the detector. As seen in Fig. 3.18 and Eq. (3.12), the line-of-centers distance $r = \ell/\cos\theta_s$ is used in the calculation of the solid angle of the detector. Combined with the projected-area falloff inherent in a Lambertian source, this yields the cosine-cubed dependence of Eq. (3.13):

$$\phi_{\text{det}} = L\,A_s\cos\theta_s\,\Omega_d = L\,A_s\cos\theta_s(A_d/r^2) \tag{3.12}$$

and

$$\phi_{\text{det}} = L\,A_s\cos\theta_s\frac{A_d}{(\ell/\cos\theta_s)^2} \propto \cos^3\theta_s. \tag{3.13}$$

3.4.4 Cosine-to-the-fourth falloff

For situations where both θ_s and θ_d are nonzero, the power transfer has a cosine-to-the-fourth falloff, considering the original perpendicular distance as the baseline. In Fig. 3.19, we consider a Lambertian source and detector that are parallel to one another but are displaced vertically ($\theta_s = \theta_d = \theta$). In this case, we find that $\phi_d \propto \cos^4\theta$, as seen in Eq. (3.15):

$$\phi_{\text{det}} = L\,A_s\cos\theta_s\,\Omega_d = L\,A_s\cos\theta_s(A_d\cos\theta_d/r^2) \tag{3.14}$$

and

$$\phi_{\text{det}} = L\,A_s\cos\theta\frac{A_d\cos\theta}{(\ell/\cos\theta)^2} \propto \cos^4\theta. \tag{3.15}$$

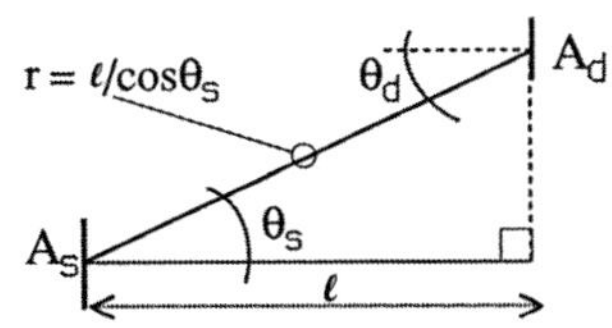

Figure 3.19 Example of cosine-to-the-fourth falloff.

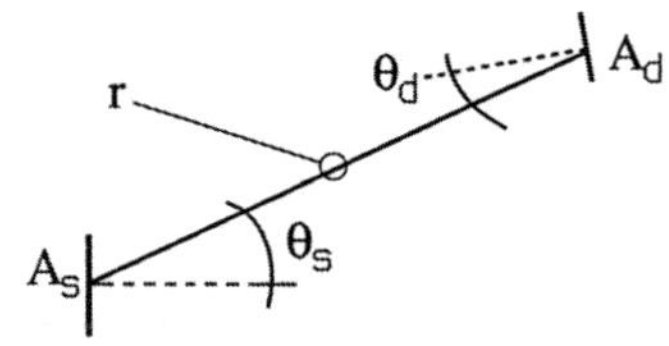

Figure 3.20 Flux transfer for an arbitrary geometry, assuming a small source and a small receiver.

For the general geometry of Fig. 3.20 involving a small Lambertian source and a small receiver, we can write the paraxial expression for the flux transfer as

$$\phi_{\rm det} = L \times A_s \cos\theta_s \times \Omega_d = L \times A_s \cos\theta_s \times \frac{A_d \cos\theta_d}{r^2}. \qquad (3.16)$$

For the paraxial Eq. (3.16) to be valid, the angles θ_s and θ_d themselves do not need to be small, but the angular subtense of the source as seen from the receiver and the angular subtense of the receiver as seen from the source must be small. This is the essence of the paraxial approximation. In other words, the line-of-centers distance r should not vary much if we integrate over the incremental areas of source and receiver to sum the incremental power transferred. We can always break a large source or large receiver into incremental areas and sum the contributions to the total flux transferred using Eq. (3.16). This approach considers the orientations of the incremental source and receiver areas with respect to each pairwise line-of-centers distance.

3.4.5 Tilting a Lambertian source

Consider the situation in Fig. 3.21, where an extended Lambertian source is viewed by the detector through an aperture. This situation maintains a constant solid angle of the source as the source is tilted. The source must be large enough to overfill the aperture so that the aperture defines Ω_s, as seen from the detector. Let us further assume that the aperture is a *cold shield* that does not contribute flux onto the detector. When the source normal is along the line of centers to the detector (no tilt), the flux reaching the detector is

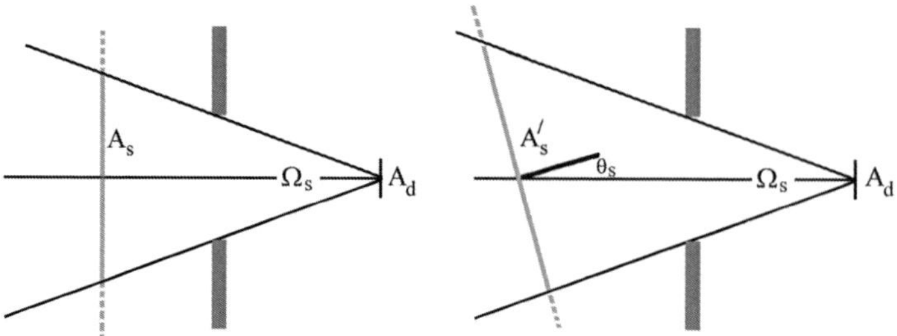

Figure 3.21 Tilting a Lambertian source, viewed through an aperture that defines a constant Ω_s.

$\phi = L\,\Omega_s A_d$. This flux can also be written as $\phi = L\,A_s\,\Omega_d$, where A_s is the area of the source that is visible to the detector through the aperture. The extent of the contributing source area A_s is defined by rays from the center of the detector through the edges of the cold shield.

When the source is tilted by an in-plane angle θ_s, a larger contributing area of the source $A_s^{\prime} = A_s/\cos\theta_s$ is visible through the aperture, but this increase in the contributing area is negated because the source is no longer parallel to the detector. The flux reaching the detector has not changed because Ω_s remains unchanged. This flux can be expressed as $\phi_{det} = L\,\Omega_s\,A_d$, or equivalently as

$$\phi = LA_s^{\prime}\cos\theta_s\,\Omega_d = L\frac{A_s}{\cos\theta_s}\cos\theta_s\,\Omega_d = L\,A_s\,\Omega_d. \tag{3.17}$$

3.4.6 Lambertian radiator: relation of exitance and radiance

For a flat Lambertian source, the following relationship holds between exitance and radiance:

$$M = \pi L \quad L = M/\pi. \tag{3.18}$$

Because, as seen in Fig. 3.22, the flat source radiates into the forward hemisphere ($\Omega = 2\pi$ sr), why isn't the relationship $M = 2\pi\,L$?

As seen in Fig. 3.17, the intensity falls off in proportion to $\cos\theta_s$. When summed over all angles in the forward hemisphere, an extra factor of ½ results:

$$M = \frac{\phi_{total}}{A_{source}} = \frac{\sum \partial^2\phi}{A_{source}} = \int_{hemisphere} L\,\cos\theta_s\partial\Omega_d = \pi\,L. \tag{3.19}$$

Performing the integration of Eq. (3.19) explicitly requires an expression for the differential solid angle, $d\Omega = dA/r^2 = (r\,\sin\theta\,d\varphi\,r\,d\theta)/r^2 = \sin\theta\,d\theta\,d\varphi$, seen in Fig. 3.23, yielding the following expression:

$$M = \frac{\phi_{total}}{A_{source}} = \int_{hemisphere} L\cos\theta_s\partial\Omega_d = \int_0^{2\pi} d\varphi \int_0^{\pi/2} L\cos\theta\sin\theta d\theta = \pi L. \tag{3.20}$$

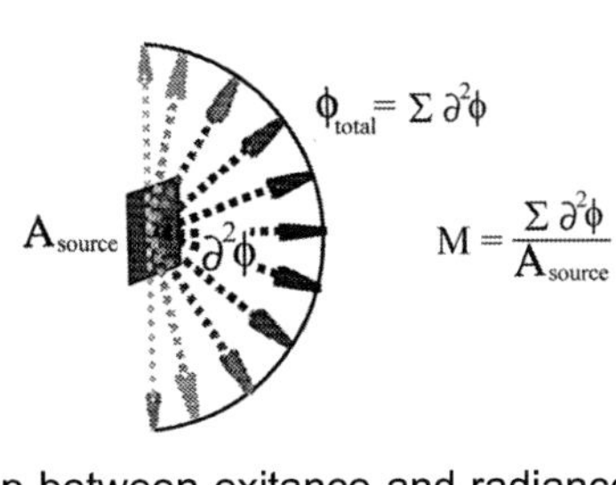

Figure 3.22 Relationship between exitance and radiance for a Lambertian source.

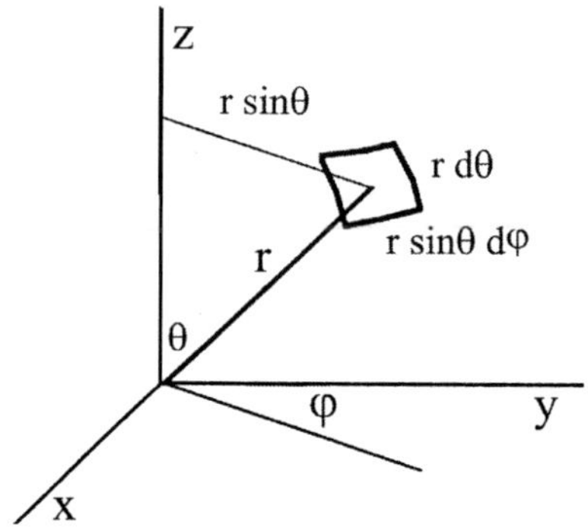

Figure 3.23 Differential solid angle.

3.5 Flux Transfer in Image-Forming Systems

We first consider the case of an extended-source object and a single thin lens with the stop at the lens, as shown in Fig. 3.24. This simple case has the same chief-ray angle in object and image space. The power ϕ collected by the system is transferred to the image plane, assuming no reflection or absorption losses in the lens. Given the definition of Eq. (2.1) for transverse magnification, the relationship for area magnification is

$$A_{\text{img}} = A_{\text{obj}}(q/p)^2. \tag{3.21}$$

Within the small-angle approximation, four equivalent $A\Omega$ products can be used to calculate ϕ, Eqs. (3.22 – 3.25):

$$\phi = L_{\text{obj}} \times A_{\text{obj}} \times \Omega_{\text{lens from obj}} = L_{\text{obj}} A_{\text{obj}} A_{\text{lens}}/p^2, \tag{3.22}$$

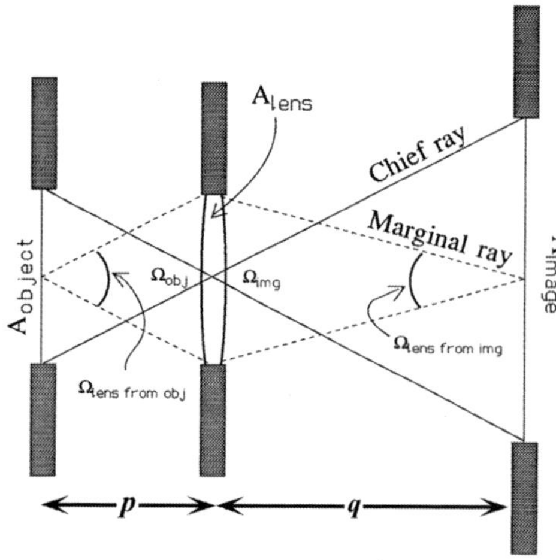

Figure 3.24 Flux transfer for a thin lens with aperture stop at the lens.

$$\phi = L_{\text{obj}} \times A_{\text{img}} \times \Omega_{\text{lens from img}} = L_{\text{obj}} A_{\text{obj}} (q/p)^2 A_{\text{lens}}/q^2, \tag{3.23}$$

$$\phi = L_{\text{obj}} \times A_{\text{lens}} \times \Omega_{\text{obj}} = L_{\text{obj}} A_{\text{lens}} A_{\text{obj}}/p^2, \tag{3.24}$$

and

$$\phi = L_{\text{obj}} \times A_{\text{lens}} \times \Omega_{\text{img}} = L_{\text{obj}} A_{\text{lens}} A_{\text{img}}/q^2. \tag{3.25}$$

Equations (3.24) and (3.25) have the interpretation that the lens acts as a secondary source of radiation with radiance equal to the radiance of the object. The $A\Omega$ product is interpreted as a flux-gathering capability, counting the ray paths from object to image, as illustrated in Fig. 3.25. Similar to the situation in Fig. 3.13 for the non-imaging case, the source radiance multiplied by the $A\Omega$ product gives the flux transferred. From Eqs. (3.22) through (3.25), we see that the power transferred from object to image is proportional to FOV^2 and $[1/(F/\#)]^2$. Recall that FOV and $F/\#$ are defined in terms of planar angles, where area and solid angle are proportional to squares of these quantities.

3.5.1 Why is radiance analogous to perceived brightness?

We now return to further consideration of radiance for Lambertian sources. It can be confusing that the *intensity* of a Lambertian source falls off in proportion to $\cos\theta_s$, as seen in Fig. 3.17, but that the *perceived brightness* of a Lambertian source is constant with respect to θ_s, as seen in Fig. 3.10. We consider the situation in Fig. 3.26. The perceived brightness of the source is proportional to the amount of power received by a single retinal photoreceptor. We use Eq. (3.24), $\phi = L_{\text{obj}} \times A_{\text{lens}} \times \Omega_{\text{obj}}$, with A_{lens} defined by the pupil

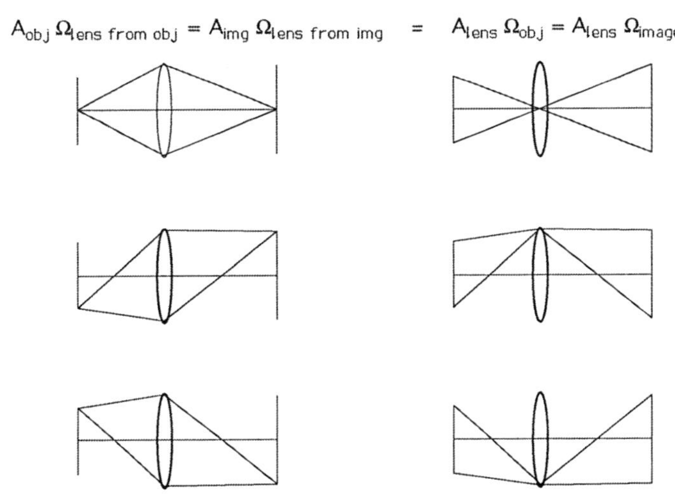

Figure 3.25 Interpretation of the $A\Omega$ product for an image-forming system.

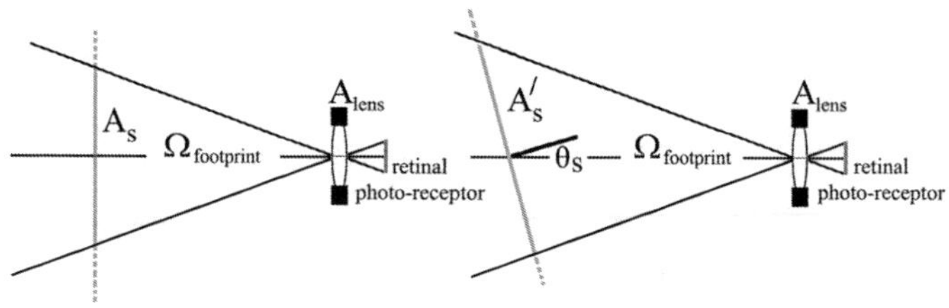

Figure 3.26 A Lambertian source has constant perceived brightness as the source tilts.

of the eye and Ω_{obj} defined by the footprint of the retinal receptor at the source plane. Similar to the situation shown in Fig. 3.21, assuming that the spatial extent of the source overfills $\Omega_{footprint}$, the amount of power received on the retinal receptor will not change if the source is tilted by an angle θ_s, as long as the source radiance L_{obj} is the same in the direction θ_s as it is at $\theta_s = 0$. This is consistent with the definition of a Lambertian source with radiance independent of view angle θ_s.

3.5.2 Example calculation of image irradiance (extended source)

In Fig. 3.27, we consider a diagram illustrating a thin-lens imaging system. Values of the relevant variables for the example calculation are: $p = 5$ cm; $q = 3$ cm; $\mathcal{M} = -q/p = -3/5$; $A_{lens} = 1$ cm^2; $L_{obj} = 2$ W/(cm^{2}sr); $A_{obj} = 1\text{ mm} \times 1\text{ mm} = 0.01\text{ cm}^2$; and $A_{img} = A_{obj} \times \mathcal{M}^2 = 0.01\text{ cm}^2 \times (-3/5)^2 = 0.0036\text{ cm}^2$.

We use Eq. (3.22) for the calculation of power transferred from object to image: $\phi = L_{obj}\ A_{obj}\ \Omega_{lens\ from\ obj} = L_{obj}\ A_{obj}\ A_{lens}/p^2$, $\phi = 2\text{ W/(cm}^2\text{ sr)} \times 0.01\text{ cm}^2 \times (1\text{ cm})^2/(5\text{ cm})^2 = 8 \times 10^{-4}$ W. When divided by the image area, this yields an image irradiance of $E_{image} = \phi/A_{image} = 8 \times 10^{-4}\text{ W}/0.0036\text{ cm}^2 = 0.22\text{ W/cm}^2$.

We consider two cases for calculating the power received by a detector in the image plane. For $A_{det} < A_{img}$, the detector acts as a field stop. For example, a 20-μm by 20-μm detector located on-axis in the image collects $\phi_{det} = E_{image} \times A_{det} = 0.22\text{ W/cm}^2 \times (20 \times 10^{-4}\text{ cm})^2 = 0.88\ \mu$W. We can use

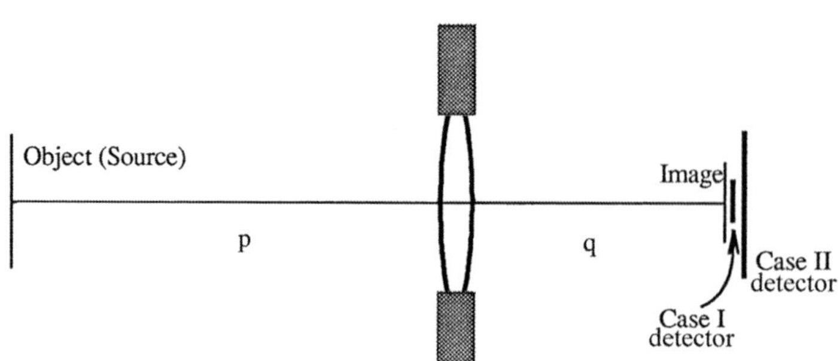

Figure 3.27 Imaging of an extended-source object.

image irradiance in this way to calculate the flux on a detector, provided that the irradiance is uniform over the detector surface. For situations where $A_{det} > A_{img}$, the source (object) acts as a field stop. For instance, for a 1-mm^2 detector located on-axis in the image, $A_{det} = 1\ mm^2 > A_{img} = (3/5\ mm)^2$. In this case, the detector collects the entire power transferred to the image plane, 8×10^{-4} W.

3.5.3 When to use a lens?

A common experimental situation is that the detector does not receive sufficient flux in a given configuration. In this case, we often consider adding a lens to the setup. Remember that the irradiance at the plane of the detector depends on the brightness of the source and the solid angle of the source (or the lens acting as a secondary source), as seen from the detector. Let us compare three situations seen in Figs. 3.28 through 3.30. In Fig. 3.28, the source and detector are at a considerable distance. In this case, the power transferred is $\phi = LA_sA_d/r^2$ and the irradiance at the detector is $E = LA_s/r^2$.

In Fig. 3.29, the source and detector are quite close. In this case, the power collected by the detector is the source exitance multiplied by the detector area, $\phi = M\ A_d = L\pi\ A_d$, and the irradiance at the detector is simply the source exitance M. In Fig. 3.30, the source is imaged onto the detector. Assuming that the detector is uniformly illuminated, the power on the detector is $\phi = L\ A_d\ \Omega_{lens}$ and the irradiance is $E = L\ \Omega_{lens}$, where Ω_{lens} is the solid angle of the lens as seen from the detector.

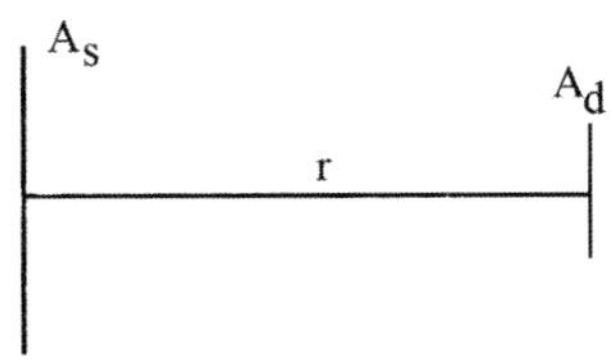

Figure 3.28 Source and detector are separated by distance *r*.

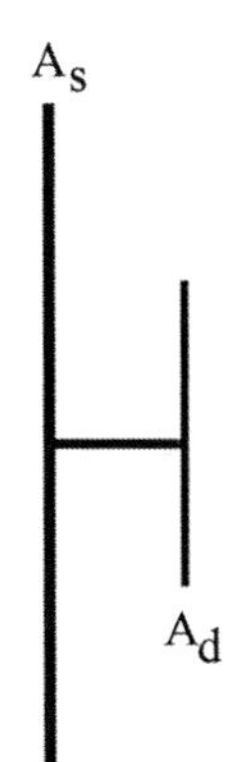

Figure 3.29 Source and detector are closely spaced.

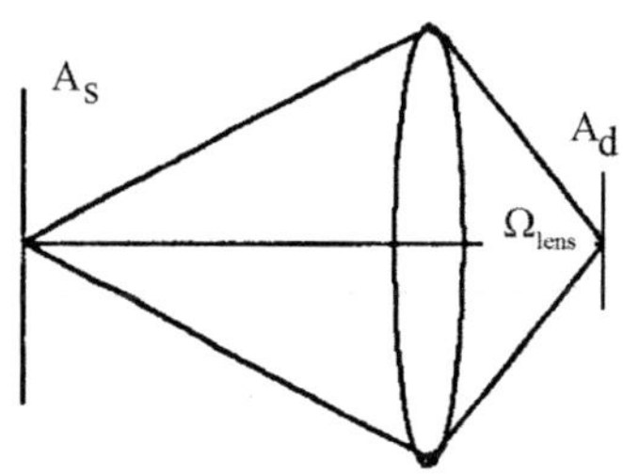

Figure 3.30 Source is imaged onto detector.

If the detector can be brought sufficiently close to the source so that the source subtends a large solid angle as seen from the detector, that is the best that can be done from a flux-transfer viewpoint. If the source and detector are constrained to be separated by a certain distance, a lens can put more flux on the detector if the angular subtense of the lens, as seen from the detector, is larger than the angular subtense of the source without the lens. This configuration implies a small image-side *F*/# of the lens. From Eq. (3.21), the image of the source is then small at the detector location, so the detector can be small also. We will see in Chapter 7 that smaller detectors have less noise, which increases SNR. There is no radiometric benefit in having a detector that is larger than the image of the source, although a larger detector may make system alignment easier.

3.5.4 Cosine-to-the-fourth falloff in imaging systems

Using Eq. (3.25), $\phi = L_{obj} \times A_{lens} \times \Omega_{img}$, the amount of flux reaching the image plane is numerically equivalent to having a source the size of the lens that has the same radiance as the original object. Equation (3.25) can be a useful starting point for some calculations. For instance, in Fig. 3.31, the chief ray makes a substantial angle θ with the optical axis. The image irradiance falls off from its on-axis value E_0 as $\cos^4\theta$, which is consistent with Fig. 3.19, with the lens acting as an intermediate source of radiance L.

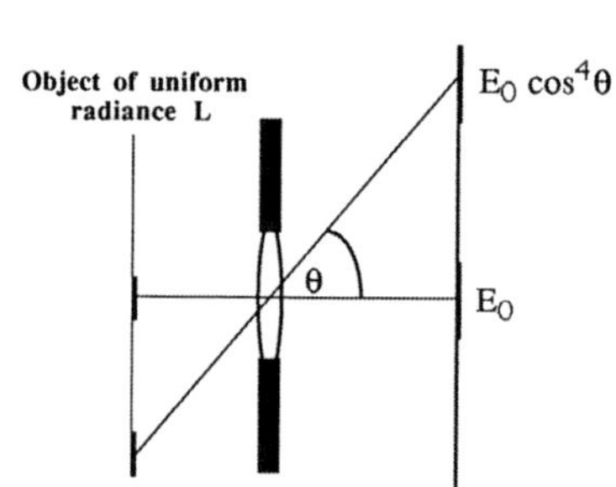

Figure 3.31 Cosine-to-the-fourth falloff in an imaging system.

3.5.5 Irradiance depends on the image-space F-number

The power ϕ collected by the lens is reformatted to form an image of the object. The image-plane irradiance can be directly found as $E_{\text{image}} = \phi/A_{\text{img}}$. Alternatively, we can use Eq. (3.25), $\phi = L_{\text{obj}}\, A_{\text{lens}}\, A_{\text{img}}/q^2$. This yields (within the paraxial approximation) $E_{\text{image}} = L_{\text{obj}}\, A_{\text{lens}}/q^2$. Using $A_{\text{lens}} = (\pi/4)(D_{\text{lens}})^2$ and $(F/\#)_{\text{image-space}} = q/D_{\text{lens}}$, we can rewrite the image-plane irradiance as

$$E_{\text{image}} = L_{\text{obj}} \frac{\pi}{4} \frac{1}{(F/\#)^2_{\text{image-space}}}. \tag{3.26}$$

When imaging an extended source, the image irradiance depends solely on the source radiance and the image-space $F/\#$, as seen in Fig. 3.32.

For a situation with a small image-space $F/\#$, the paraxial assumptions are not valid, and the general result is

$$E_{\text{image}} = \frac{\pi L_{\text{obj}}}{1 + 4(F/\#)^2_{\text{image-space}}}. \tag{3.27}$$

We begin the non-paraxial development with Eq. (3.8), rewritten as

$$\partial^2 \phi_{\text{d}} = L\, \partial A_{\text{s}} \cos\theta_{\text{s}}\, \partial\Omega_{\text{d}} = L\, \partial A_{\text{s}} \cos\theta_{\text{s}} \frac{\partial A_{\text{d}} \cos\theta_{\text{d}}}{r^2}. \tag{3.28}$$

We then integrate over the source and detector, with the variable r representing the line-of-centers distance between the incremental source and detector areas,

$$\phi_{\text{d}} = \int_{A_{\text{s}}} \int_{A_{\text{d}}} L\, \mathrm{d}A_{\text{s}} \cos\theta_{\text{s}} \frac{\cos\theta_{\text{d}}\, \mathrm{d}A_{\text{d}}}{r^2}, \tag{3.29}$$

as shown in Fig. 3.33. We identify the following relationships in this figure: $\theta_{\text{s}} = \theta_{\text{d}} \equiv \theta$, $\mathrm{d}A_{\text{s}} = 2\pi R\, \mathrm{d}R$, $R = z\ \tan\theta$; $\mathrm{d}R = (z/\cos^2\theta)\ \mathrm{d}\theta$; and $\mathrm{d}A_{\text{s}} = 2\pi\ (z\ \tan\theta)\ z\ (\mathrm{d}\theta/\cos^2\theta)$. With these substitutions, we find that

$$\phi_{\text{d}} = LA_{\text{d}} 2\pi \int_0^{\theta_{\max}} \cos\theta\ \sin\theta\, d\theta = A_{\text{d}} \pi L \sin^2\theta_{\max} \tag{3.30}$$

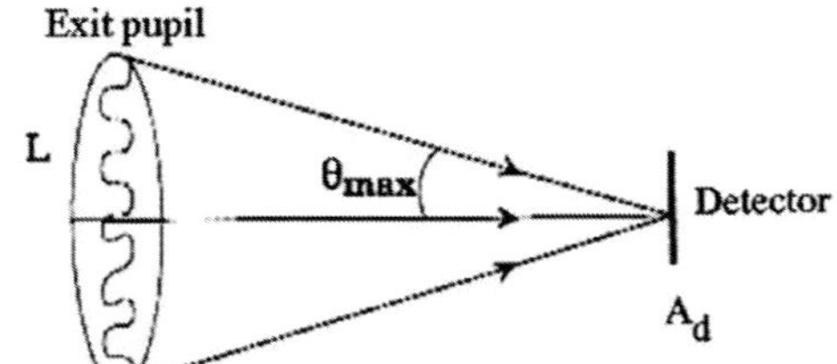

Figure 3.32 Irradiance is proportional to the inverse square of the image-space $F/\#$, in the paraxial limit.

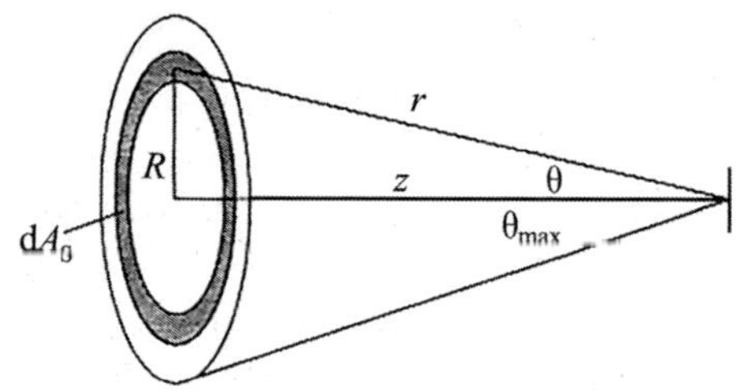

Figure 3.33 Calculation of image irradiance in the non-paraxial case.

and

$$E = \frac{\phi_d}{A_d} = \pi L \sin^2\theta_{max}. \tag{3.31}$$

By the construction shown in Fig. 3.34, we express the image irradiance in terms of the image-space *F/#* as follows:

$$\sin\theta_{max} = \frac{D/2}{\sqrt{f^2 + \left(\frac{D}{2}\right)^2}} \qquad \sin^2\theta_{max} = \frac{1}{4(F/\#)^2 + 1}. \tag{3.32}$$

These substitutions yield a final expression for image-plane irradiance in the non-paraxial case, consistent with Eq. (3.27):

$$E = \pi L \sin^2\theta_{max} = \frac{\pi L_{obj}}{1 + 4\,(F/\#)^2_{image\text{-}space}}. \tag{3.33}$$

If the source and detector are close enough that $\theta_{max} \sim 90$ deg, then $E_{det} = \pi L = M_{source}$, as seen in Fig. 3.29.

3.5.6 Aperture stop not at the lens

For a general optical system as seen in Fig. 3.35, where the aperture stop is not located at the lens, the image of the aperture stop (ASTOP) viewed from the object plane is the entrance pupil (ENP) and the image of the aperture stop (ASTOP) viewed from the image plane is the exit pupil (EXP). In this

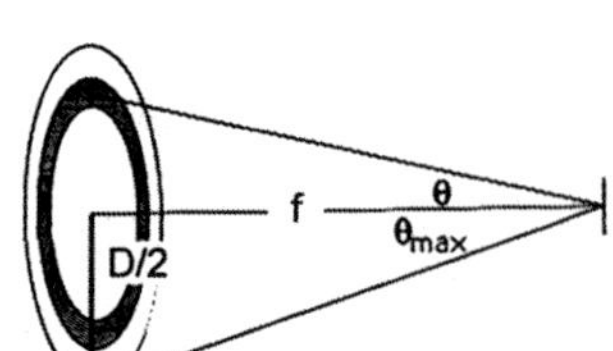

Figure 3.34 Relating θ_{max} to image-space *F/#*.

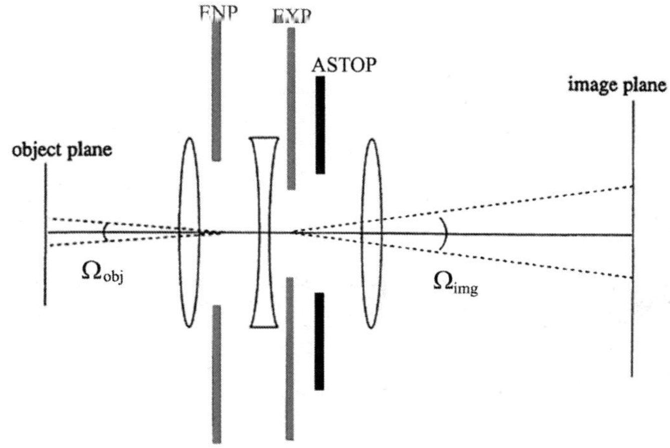

Figure 3.35 For a general optical system, the chief-ray angle may be different in object space and image space.

general case, the chief ray will have a different angle in object space and image space if the ENP and EXP diameters are different. The following relationships apply because of equivalence of the $A\Omega$ products, as seen in Eqs. (3.22) to (3.25):

$$A_{\text{ENP}} \times \Omega_{\text{obj}} = A_{\text{EXP}} \times \Omega_{\text{img}}. \tag{3.34}$$

3.5.7 Radiance of an extended source is conserved

Assuming air for both the object-space and image-space media, the radiance in object and image planes is equal: $L_{\text{img}} = L_{\text{obj}}$. This equality is a consequence of Eqs. (3.22) to (3.25), which assumes that reflection or absorption losses in the lenses are negligible. These relationships also assume paraxial geometrical optics as in Eq. (2.1) so that a point in the object maps to another point (not a blur spot) in the image and an area in the object plane maps to another area in the image plane, as seen in Fig. 3.24. Thus, effects of diffraction and aberrations are also assumed to be small. With reference to Fig. 3.36, we see that

$$A_1\Omega_1 = A_{\text{obj}}\frac{A_{\text{lens}}}{p^2} = A_{\text{img}}\frac{A_{\text{lens}}}{q^2} = A_{\text{obj}}\left(\frac{q^2}{p^2}\right)\frac{A_{\text{lens}}}{q^2} = A_2\Omega_2. \tag{3.35}$$

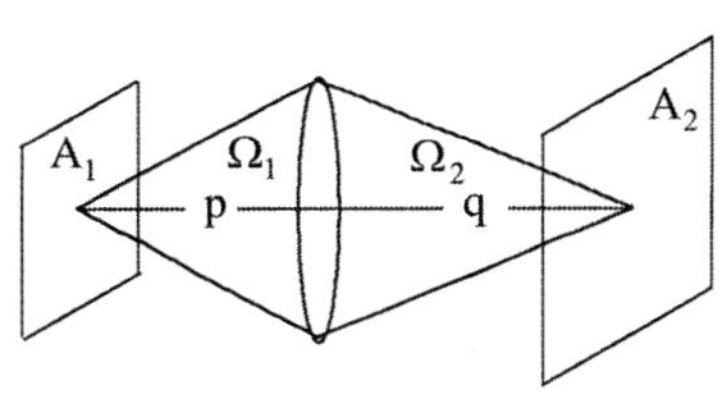

Figure 3.36 Illustration of the conservation of radiance.

The power transmitted through the lens can be written as $\phi_{\text{thru-lens}} = L_{\text{obj}} A_1 \Omega_1$ so that $L_{\text{obj}} = \phi_{\text{thru-lens}}/(A_1\Omega_1)$. The image-plane radiance $L_{\text{img}} = \phi_{\text{thru-lens}}/(A_2\Omega_2)$. By the equality of the $A\Omega$ products, the image-plane radiance is equal to the object-plane radiance under the assumptions noted.

3.5.8 Dilution of radiance by a finite impulse response

A finite-sized impulse response reduces the radiance of the image by spreading the image-plane flux into a region larger than indicated by geometrical optics (caused by either aberrations or diffraction). A point in the object no longer maps to a point in the image. The area–magnification relationship, $A_{\text{img}} = A_{\text{obj}}\ (q/p)^2$, of Eq. (3.21) no longer holds because the image area is convolved in two dimensions with the impulse response, as in Eq. (2.4). Hence, the area of the image is increased, but the flux through the lens is the same. The implication is that the image-plane radiance is reduced from the case of paraxial geometrical optics:

$$L_{\text{img-plane}}\downarrow = \phi_{\text{thru-lens}}/(A_2\uparrow\Omega_2). \tag{3.36}$$

For example, consider the system illustrated in Fig. 3.37. The diffraction spot projected to the object plane is larger than the dimension of the object. The image of this source is the convolution of the geometrical image of the source and the image-plane diffraction spot. This image irradiance distribution is of larger spatial extent than the geometrical image, thereby reducing the image-plane radiance below that of the object-plane radiance.

3.6 Point-Source Imaging

Often the object of interest is effectively a point source, one with a size much smaller than the resolution spot projected to the object plane, an unresolved object. The point source is specified in terms of its intensity I (W/sr). With reference to Figs. 3.6 and 3.38, the total power collected is $\phi = I \times \Omega_{\text{lens}} = \text{I} \times A_{\text{lens}}/p^2$.

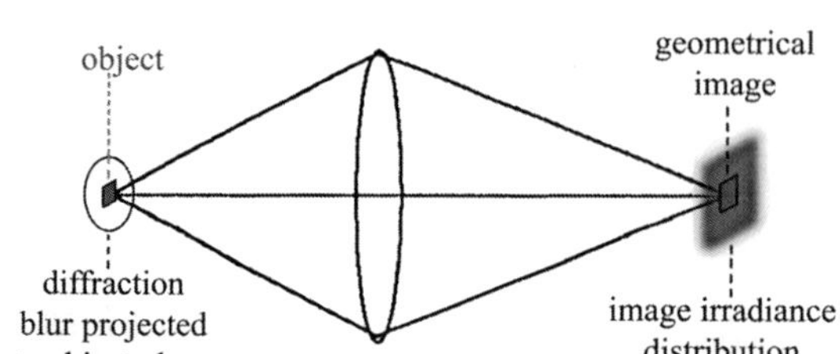

Figure 3.37 Image-plane radiance is reduced by a finite-size impulse response.

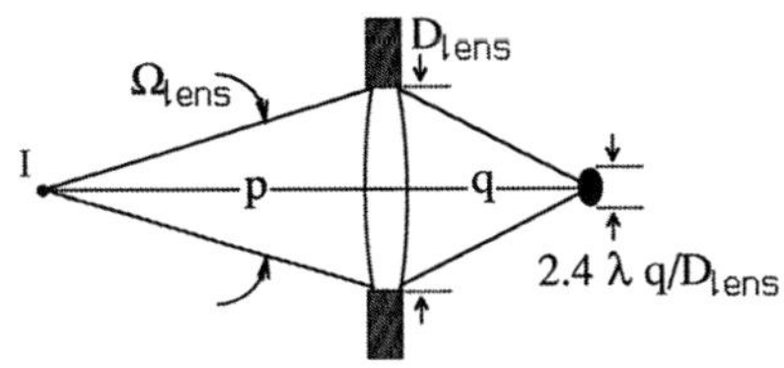

Figure 3.38 Calculation of image-plane irradiance for an unresolved object.

If the lens is diffraction-limited, 84% of ϕ is concentrated into a spot of diameter 2.4 λ $(F/\#)_{\text{image-space}}$. In that case, we can approximate the on-axis irradiance in the image plane as

$$E_{\text{img}} = \frac{\text{power in spot}}{\text{area of spot}} \approx \frac{0.84\, I\, \Omega_{\text{lens}}}{\frac{\pi}{4}(2.4\lambda(F/\#)_{\text{image-space}})^2}. \tag{3.37}$$

Bibliography

McCluney, W. R., *Introduction to Radiometry and Photometry, 2nd Edition*, Artech House, Boston (2014).

Wolfe, W. L., *Introduction to Radiometry*, SPIE Press, Bellingham, Washington (1998) [doi: 10.1117/3.287476].

Chapter 4
Thermal Sources

4.1 Introduction

Heated solid bodies emit radiation that is particularly concentrated in the infrared region of the spectrum, roughly from 1 to 10 μm in wavelength. Such sources emit their radiation in a continuum of wavelengths rather than at discrete spectral lines characteristic of gaseous emission. To describe the radiation characteristics of a source that emits a finite total power over a range of wavelengths, we must introduce spectral quantities. Spectral radiometric quantities (Table 4.1) have units of microns in the denominator and are denoted with a subscript λ to distinguish them from the corresponding integrated *in-band* quantities.

For example, spectral radiance L_λ is the radiance per micron of wavelength interval. For a measurement over a 1-μm spectral bandpass, the integrated radiance numerically equals the spectral radiance. For a spectral bandpass larger or smaller than 1 μm, the in-band radiometric quantity scales with the width of the spectral bandpass $\Delta\lambda = \lambda_1 - \lambda_2$, as seen in Eqs. (4.1) and (4.2):

$$M\left[\frac{\mathrm{W}}{\mathrm{cm}^2}\right] = \int_{\lambda_1}^{\lambda_2} M_\lambda(\lambda)\left[\frac{\mathrm{W}}{\mathrm{cm}^2\,\mu\mathrm{m}}\right] \mathrm{d}\lambda[\mu\mathrm{m}] \tag{4.1}$$

and

$$L\left[\frac{\mathrm{W}}{\mathrm{cm}^2\ \mathrm{sr}}\right] = \int_{\lambda_1}^{\lambda_2} L_\lambda(\lambda)\left[\frac{\mathrm{W}}{\mathrm{cm}^2\ \mathrm{sr}\ \mu\mathrm{m}}\right] \mathrm{d}\lambda[\mu\mathrm{m}]. \tag{4.2}$$

Table 4.1 Spectral radiometric quantities.

Spectral radiance: $L_\lambda \dfrac{\mathrm{W}}{\mathrm{cm}^2\ \mathrm{sr}\ \mu\mathrm{m}}$
Spectral exitance: $M_\lambda \dfrac{\mathrm{W}}{\mathrm{cm}^2\ \mu\mathrm{m}}$
Spectral flux: $\phi_\lambda \dfrac{\mathrm{W}}{\mu\mathrm{m}}$

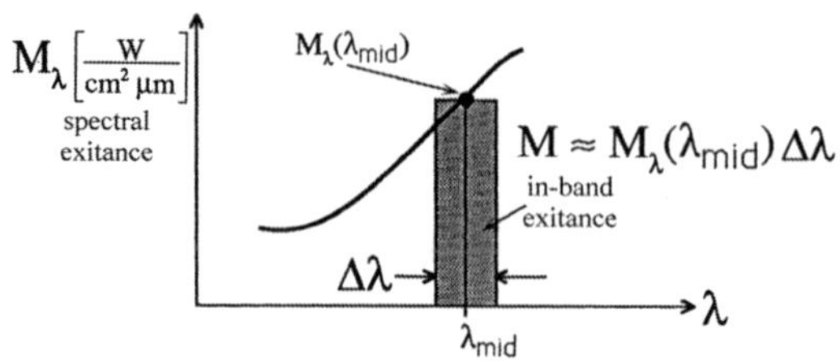

Figure 4.1 Trapezoidal integration.

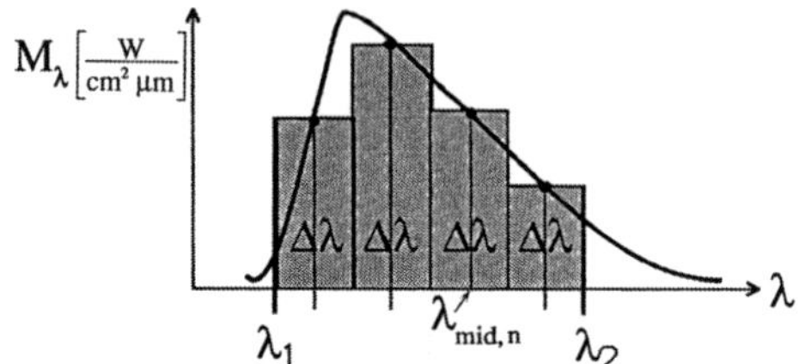

$$M = \int_{\lambda_1}^{\lambda_2} M_\lambda(\lambda)\, d\lambda \approx \Delta\lambda \sum_n M_\lambda(\lambda_{mid,n})$$

Figure 4.2 Integration over a wider passband.

The integration is usually carried out numerically using the trapezoidal approximation seen in Fig. 4.1. This process can be made as accurate as desired, by choice of the spectral interval size $\Delta\lambda$. As seen in Fig. 4.2, to maintain accuracy for a wider passband, individual intervals are summed together. The spectral dependence of the radiation function being integrated should be considered when choosing the interval widths and locations.

4.2 Blackbody Radiation

Radiation from an ideal blackbody is entirely determined by the blackbody temperature. A blackbody has the maximum spectral exitance possible for a heated body at a particular specified temperature. This condition is true over any $\Delta\lambda$ interval and also over all wavelengths. A blackbody is a convenient baseline for comparison with real sources because any real source at a given temperature is constrained to emit less radiation than a blackbody source at the same temperature. Because of the population inversion inherent in laser gain media, the usual thermodynamic definition of temperature does not apply. Thus, lasers can have spectral radiances that far exceed any blackbody source.

Blackbody radiation is cavity radiation. The radiation emitted from a small aperture in front of an isothermal cavity depends solely on the temperature of the cavity. The spectral radiance does not depend on the cavity material. Figure 4.3 shows typical cavity configurations that are the basis for construction of practical blackbody simulators.

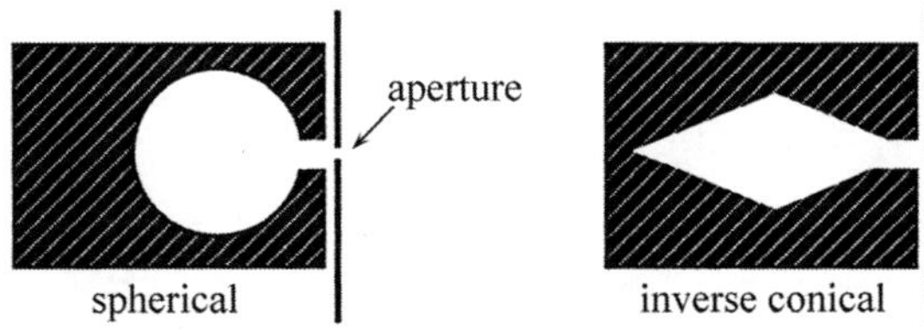

Figure 4.3 Cavity radiator configurations.

Figure 4.4 Blackbody simulator.

Figure 4.4 shows a blackbody simulator used for laboratory calibration that is constructed as a heated enclosure with a small aperture limiting the effective radiating area. Electrical heating coils are wound around a cavity of fire-brick material. A selectable aperture wheel defines the radiating area A_s.

To be an accurate blackbody simulator, the active area of the source must absorb any incident light. Kirchhoff's law (Section 4.5.1) states that a perfect absorber is a perfect emitter. The cavity configuration traps any incident radiation regardless of whether the cavity material has unit absorption. In Fig. 4.5 we see a conical blackbody configuration. Assume that each time a ray hits the cavity wall, half of the power in the ray is absorbed and half is reflected. The cavity walls are shaped so that any ray entering the aperture is

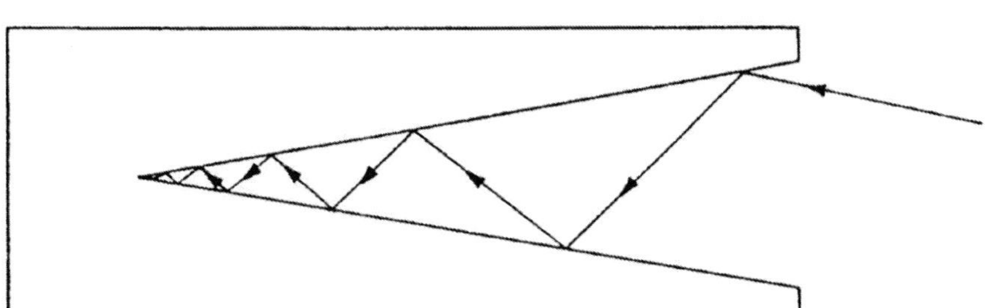

Figure 4.5 Cavity enforces unit absorption in a conical blackbody.

reflected many times inside the cavity before the ray has a chance to escape the cavity. After two reflections, 75% of the power in the ray is absorbed. After three reflections, 87.5% of the power is absorbed. After ten reflections, more than 99% of the power is absorbed. Light is not allowed to escape the cavity, and hence the aperture/cavity configuration has unit absorption, as required.

4.3 Planck's Equation

The spectral exitance of a blackbody is a function of both temperature and wavelength. For a given temperature, exitance is a function of wavelength. This results in a family of spectral-exitance curves parameterized on temperature. The equation for spectral exitance is Planck's equation, Eq. (4.3),

$$M_\lambda(\lambda, T) = \frac{2\pi hc^2}{\lambda^5[e^{hc/\lambda kT} - 1]} \left[\frac{\text{W}}{\text{cm}^2\ \mu\text{m}}\right], \tag{4.3}$$

where Planck's constant $h = 6.6 \times 10^{-34}$ J sec, speed of light $c = 3 \times 10^8$ m/sec, Boltzmann's constant $k = 1.38 \times 10^{-23}$ J/K, and T is the absolute temperature specified in Kelvin (°C + 273). Room temperature is about 300 K. Figure 4.6 shows a family of Planck's curves, covering a range of temperatures from 600 K to 2000 K.

Two characteristics can be noted directly from Fig. 4.6. As the source temperature increases, the total area under the curve (integrated exitance) increases much faster than the absolute temperature. This relates to the Stefan–Boltzmann law (Section 4.3.2). As the source temperature increases, the wavelength of peak exitance moves toward shorter wavelengths. This relationship is known as Wien's displacement law (Section 4.3.3).

4.3.1 Other forms of Planck's equation

Because blackbodies are Lambertian by definition,[1] radiance is exitance divided by π, as seen in Eq. (3.18). Starting with the formula for spectral exitance, Eq. (4.3), we find for spectral radiance (in energy-based units)

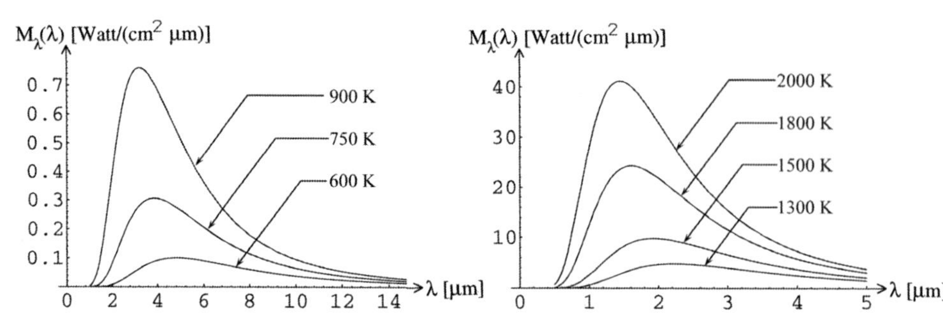

Figure 4.6 Blackbody spectral exitance curves for 600 K < T < 2000 K.

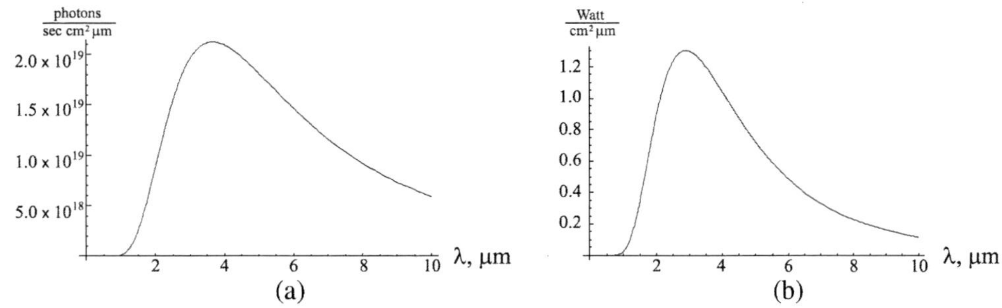

Figure 4.7 (a) Photon exitance [photon/(sec cm^2 μm)] vs λ [μm] for a 1000 K blackbody. (b) Radiant exitance [W/(cm^2 μm)] vs λ [μm] for a 1000 K blackbody.

$$L_\lambda(\lambda, T) = \frac{2hc^2}{\lambda^5[e^{hc/\lambda kT} - 1]} \left[\frac{\mathrm{W}}{\mathrm{cm}^2\ \mathrm{sr}\ \mu\mathrm{m}}\right]. \tag{4.4}$$

Sometimes when calculating the response of a photon sensor, it is convenient to have Planck's equation in photon units. Because the conversion factor of Eq. (3.3) ($\mathcal{E} = hc/\lambda$ [J/photon]) between the two sets of units is wavelength dependent, the plots for spectral photon exitance are different functions of wavelength (different shape, peak wavelength, etc.) than the plots for spectral radiant exitance. We find the following expressions for spectral photon exitance and spectral photon radiance:

$$M_{\lambda,q}(\lambda, T) = \frac{2\pi c}{\lambda^4[e^{hc/\lambda kT} - 1]} \left[\frac{\text{photons}}{\text{sec cm}^2\ \mu\text{m}}\right] \tag{4.5}$$

and

$$L_{\lambda,q}(\lambda, T) = \frac{2c}{\lambda^4[e^{hc/\lambda kT} - 1]} \left[\frac{\text{photons}}{\text{sec cm}^2\ \text{sr}\ \mu\text{m}}\right]. \tag{4.6}$$

In Fig. 4.7, we compare the plots of $M_{\lambda,q}(\lambda)$ and $M_\lambda(\lambda)$ for a 1000 K blackbody, verifying that the respective Wien's laws (Section 4.3.3) are satisfied in terms of peak-emission wavelength.

4.3.2 Stefan–Boltzmann law

The Stefan–Boltzmann law relates total blackbody exitance at all λs to the source temperature, integrating out the wavelength dependence of Planck's equation seen here for radiant exitance:

$$M(T) = \int_0^\infty M_\lambda(\lambda, T)\, \mathrm{d}\lambda = \int_0^\infty \frac{2\pi hc^2}{\lambda^5[e^{hc/\lambda kT} - 1]}\, \mathrm{d}\lambda = \sigma T^4. \tag{4.7}$$

The total exitance of a blackbody source (and, multiplying by source area, the total power radiated) increases as the fourth power of the absolute temperature when the integration is taken over all wavelengths. The Stefan–Boltzmann constant $\sigma = 5.7 \times 10^{-12}$ W/(cm^2 K^4). A body at room temperature (300 K) has a total exitance of 4.6×10^{-2} W/cm^2. Doubling the absolute temperature of the body to 600 K increases the total exitance 16-fold to 0.74 W/cm^2. There is also a Stefan–Boltzmann law for photon flux:

$$M_q(T) = \int_0^\infty M_{q,\lambda}(\lambda, T)\,\mathrm{d}\lambda = \int_0^\infty \frac{2\pi c}{\lambda^4[e^{hc/\lambda kT} - 1]}\,\mathrm{d}\lambda = \sigma_q T^3, \tag{4.8}$$

where $\sigma_q = 1.5 \times 10^{11}$ photons/(sec cm^2 K^3).

However, the Stefan–Boltzmann law only holds for exitance integrated from an interval of 0 to ∞ on wavelength. We usually have a finite passband somewhere in an imaging system (e.g., atmospheric transmission, optics transmission, and spectral response of the detector). The assumption of an unlimited passband is more relevant for heat-transfer calculations than for imaging scenarios. When the range of integration is limited, the exponent of the dependence of the in-band exitance on temperature can be greater or less than 4 because, as seen in Fig. 4.6, some portions of the spectral-exitance curve change faster than T^4 (on the short-λ side of the peak) and some portions (at long wavelengths) change more slowly. We can solve numerically for the exponent using the following:

$$\frac{\int_{\lambda_1}^{\lambda_2} M_\lambda(\lambda, T + \Delta T)\,\mathrm{d}\lambda}{\int_{\lambda_1}^{\lambda_2} M_\lambda(\lambda, T)\,\mathrm{d}\lambda} = \left(\frac{T + \Delta T}{T}\right)^\gamma. \tag{4.9}$$

The exponent γ varies depending on where the system passband is located with respect to the peak of the blackbody curve.

4.3.3 Wien's displacement law

We saw in Fig. 4.6 that as the blackbody temperature increases, the peak-exitance wavelength decreases, which shows Wien's displacement law. Wien's law can be derived from the condition for the peak of the spectral-exitance function either analytically by setting $\partial M_\lambda(\lambda,T)/\partial\lambda = 0$ or by plotting the curve and noting where it peaks. Either of these approaches yields a constraint on the wavelength of maximum exitance λ_{max}:

$$\lambda_{max}[\mu\mathrm{m}] = \frac{2898}{T\,[\mathrm{K}]}, \tag{4.10}$$

typically approximated as $\lambda_{max} T \sim 3000$ μm K. The plot of λ_{max} vs blackbody temperature is a hyperbola. For example, a blackbody source at $T = 300$ K (room temperature) has its maximum exitance at ≈ 10 μm. For a 1000 K blackbody, $\lambda_{max} \approx 3$ μm. The sun approximates a blackbody at 6000 K, whose spectral radiance peaks near $\lambda_{max} \approx 0.5$ μm. There is also a Wien's law for photon flux. The condition $\partial M_{\lambda,q}(\lambda,T)/\partial\lambda = 0$ yields $\lambda_{max}T = 3670$ μm K, as shown in Fig. 4.7(a).

4.3.4 Half-power points

Additional insight to the behavior of blackbody sources can be found from the power relationships relative to the half-power points of Planck's function. We define the half-power point on the short-wavelength side of the peak as λ_{short} and the corresponding half-power point on the long-wavelength side as λ_{long}. The following relationships hold (for radiant-flux curves):

$$\lambda_{short} \sim \frac{1800\,[\mu\mathrm{m\,K}]}{T\,[\mathrm{K}]} \text{ and } \lambda_{long} \sim \frac{5100\,[\mu\mathrm{m\,K}]}{T\,[\mathrm{K}]}. \tag{4.11}$$

For example, the 300 K blackbody curve seen in Fig. 4.8 has its half-power points at 6 μm and 17 μm. The area under Planck's curve from $\lambda = 0$ to λ_{short} is 3% of σT^4. The area from λ_{short} to λ_{max} is 25% of σT^4. The area from λ_{max} to λ_{long} is 35% of σT^4, and the area from λ_{long} to ∞ is 37% of σT^4. Planck's function increases quickly at short wavelengths and slowly decreases at long wavelengths. A blackbody has 60% of its output between the short- and long-wavelength half-power points.

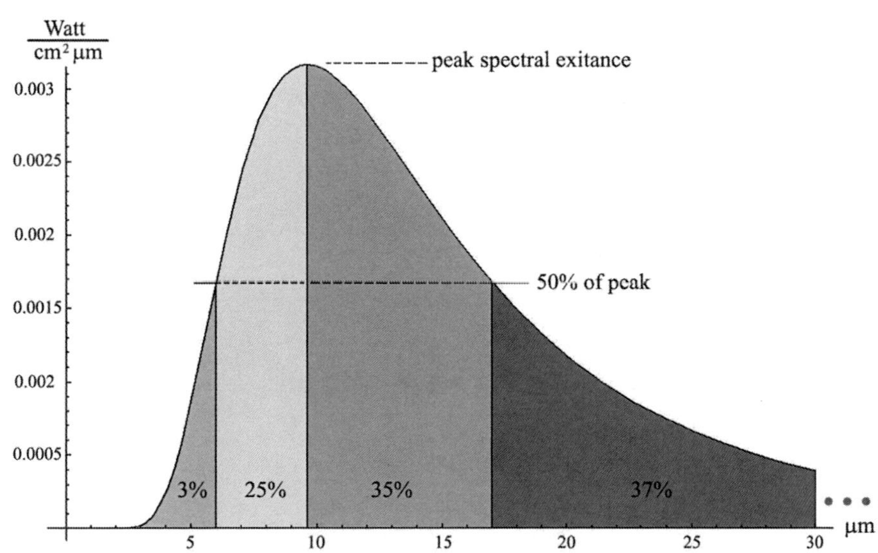

Figure 4.8 Half-power points for Planck's function at 300 K.

4.3.5 Calculations

Calculations can be made directly from Eq. (4.3). Calculation algorithms should be verified using the plots in Fig. 4.6, taking care that the correct units (W cm^{-2} μm^{-1}) are obtained. As an example, we will calculate $M_\lambda(\lambda = 3\ \mu m, T = 900\ K)$. Checking Fig. 4.9, we expect that our calculation will yield approximately $M_\lambda = 0.75$ W cm^{-2} μm^{-1}.

Considering the units in Planck's equation, the factor $(hc/\lambda kT)$ must be dimensionless because it is in the exponent. Given that this expression has units of $[\frac{hc}{\lambda kT}] = \frac{J s \times m/s}{m \times J/K \times K}$, we must input the wavelength in units of meters. The units of hc^2/λ^5 are then $[\frac{hc^2}{\lambda^5}] = \frac{J s \times m^2/s^2}{m^5} = \frac{W}{m^3}$. Planck's equation is interpreted as a spatial power density per unit wavelength interval (not a volumetric power density). To convert the units of M_λ to W cm^{-2} μm^{-1}, we multiply by a factor of 10^{-10} after the calculation is performed. Evaluating M_λ(3 μm, 900 K), we find that $hc/\lambda kT = 5.31$ and $(\exp\{5.31\} - 1) = 201.35$. Thus, $M_\lambda = 7.6 \times 10^9$ W/m^3 = 0.76 W cm^{-2} μm^{-1}, in good agreement with our initial estimate from Fig. 4.9.

Another useful calculation is the *in-band* flux over a given wavelength interval. We begin with a blackbody source at 750 K with source area $A_s = 1\ cm^2$. At a distance of 1 m, we place a 1 mm^2 detector. To calculate power on the detector within the 8- to 10-μm band, we begin with the standard flux-transfer expression, Eq. (3.10) ($\phi = L A_s \Omega_d$). This relationship also holds for spectral quantities:

$$\phi_\lambda = L_\lambda A_s \Omega_d. \tag{4.12}$$

Using Eq. (3.18), ($L = M/\pi$) for a Lambertian source:

$$\phi_\lambda = \frac{M_\lambda}{\pi} A_s \Omega_d. \tag{4.13}$$

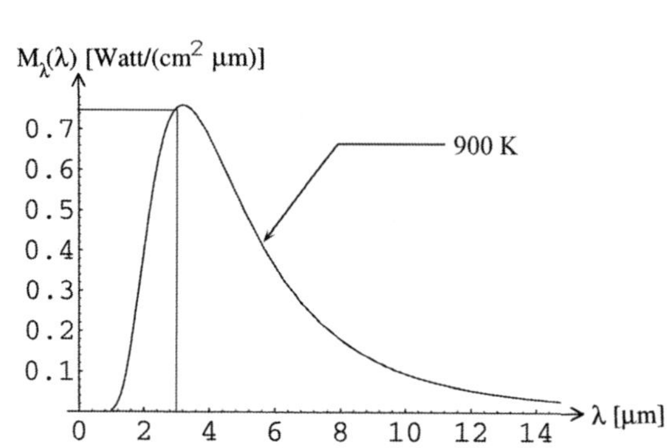

Figure 4.9 Planck's function at 900 K.

Using a one-interval integration as in Fig. 4.1, we express the in-band flux as the product of the mid-band spectral flux $\phi_{\lambda,\text{mid}}$ and the spectral bandwidth $\Delta\lambda$, which yields Eq. (4.14):

$$\phi \approx \phi_{\lambda,\text{mid}}\Delta\lambda = \frac{M_{\lambda,\text{mid}}\Delta\lambda}{\pi} A_s \Omega_d. \tag{4.14}$$

Using Eq. (4.3) for a blackbody at 750 K, we find for the mid-band spectral flux: $M_{\lambda,\text{mid}}(9\ \mu\text{m}, 750\ \text{K}) \approx 0.08 \left[\frac{\text{W}}{\text{cm}^2\ \mu\text{m}}\right]$. Equation (4.14) yields a value for in-band power on the detector of $\phi \approx \frac{0.08 \frac{\text{W}}{\text{cm}^2 \mu\text{m}} 2\ \mu\text{m}}{\pi\ \text{sr}} 1\ \text{cm}^2 \frac{(10^{-3}\ \text{m})^2}{1\ \text{m}^2} = 50\ \text{nW}$.

4.3.6 Planck's equation in terms of other spectral variables

Planck's equation of Eq. (4.3) can be expressed in variables other than wavelength, such as frequency f or wavenumber ($\tilde{\nu} \equiv 1/\lambda$). Along with a direct substitution to convert the variables, be sure to also convert the differential. To convert the expression for $M_\lambda(\lambda,T)$ [W/(cm^2 μm)] to $M_f(f,T)$ [W/(cm^2 Hz)], we include the differentials to start from [W/cm^2] quantities on both sides of the equation:

$$M_\lambda(\lambda, T)\ [\text{W}/(\text{cm}^2\mu\text{m})]\ \text{d}\lambda\ [\mu\text{m}] = M_f(f, T)[\text{W}/(\text{cm}^2\ \text{Hz})]\ \text{d}f\ [\text{Hz}]. \tag{4.15}$$

From $\lambda f = c$, we obtain by derivative an expression for $\text{d}\lambda$ in terms of $\text{d}f$:

$$|\text{d}\lambda| = |df|\frac{c}{f^2}. \tag{4.16}$$

Substituting Eqs. (4.3) and (4.16) into Eq. (4.15),

$$M_\lambda(\lambda, T)\text{d}\lambda = \frac{2\pi hc^2}{\lambda^5[e^{hc/\lambda kT} - 1]}\text{d}\lambda = \frac{2\pi hc^2}{\left(\frac{c}{f}\right)^5[e^{hf/kT} - 1]}\frac{c}{f^2}|df|. \tag{4.17}$$

We then obtain the spectral-exitance expression in terms of frequency:

$$M_f(f, T) = \frac{2\pi hf^3}{c^2[e^{hf/kT} - 1]}, \tag{4.18}$$

with units (after the factor-of-ten conversion seen in Section 4.3.5) of [W/(cm^2 Hz)].

4.4 Exitance Contrast

In the case of ground-based IR systems, the target and background are often of similar temperature. The target then has low contrast with respect to the background. To optimize target detectability, the proper choice of spectral

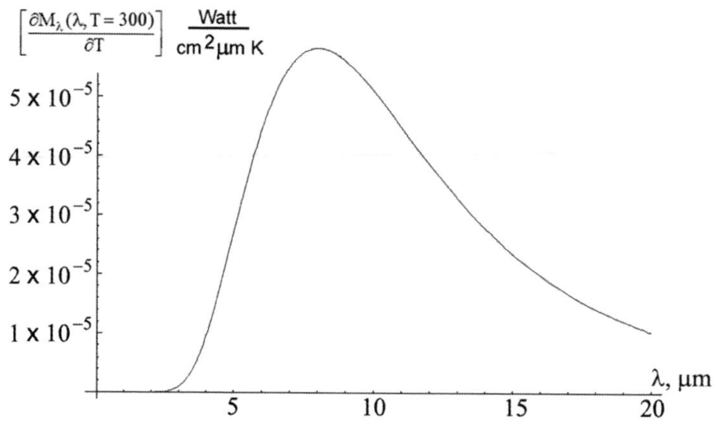

Figure 4.10 Exitance contrast $\partial M_\lambda(\lambda,T)/\partial T$ vs λ for a 300 K blackbody.

passband can be important. With plenty of flux arriving from the background, it may be beneficial to place the passband of the system in the spectral region where the exitance changes the most with temperature instead of the region of maximum exitance as defined by Wien's law. This choice of passband location will emphasize the small temperature differences between the target and the background. The wavelength of maximum exitance contrast can be found from the following equation:

$$\frac{\partial}{\partial\lambda}\left[\frac{\partial M_\lambda(\lambda,T)}{\partial T}\right]=0. \tag{4.19}$$

Solving Eq. (4.19) produces a condition of similar form to Wien's law, where

$$\lambda_{\text{max-contrast}}[\mu\text{m}]=\frac{2410}{T\,[\text{K}]}\qquad \lambda_{\text{max-contrast}}\,T=2410\,\mu\text{m K}. \tag{4.20}$$

For a blackbody of a given temperature, the maximum value of exitance contrast will occur at a shorter wavelength than the wavelength for peak exitance. For a 300 K blackbody, the peak spectral exitance occurs around 10 μm, whereas the peak exitance contrast occurs around 8 μm.

Sometimes numerical calculations must be made with exitance contrast, for instance to convert between D^* and NETD (Section 9.4.1). We can do this by computing two Planck's curves with a 1 K temperature difference and subtracting them. This is a convenient way to calculate $\partial M_\lambda(\lambda,T)/\partial T$. The function can be integrated, if needed, over any given passband. A plot of $\partial M_\lambda(\lambda,T)/\partial T$ vs λ is shown in Fig. 4.10 for a 300 K blackbody.

4.5 Emissivity

Although the assumption of blackbody radiation provides an initial model for sources, a correction factor is often applied to improve the accuracy of flux-transfer calculations. This factor is the emissivity ε, a dimensionless number that is the ratio of the radiation from the actual source to the radiation from a blackbody at the same temperature. Emissivity characterizes how the radiation spectrum of a source differs from that of a blackbody. A commonly used definition is that of the hemispherical emissivity, written in terms of the ratio of exitance quantities:

$$\varepsilon(\lambda, T) = \frac{M_{\lambda,\,\mathrm{source}}(\lambda, T)}{M_{\lambda,\,\mathrm{BB}}(\lambda, T)} \leq 1. \tag{4.21}$$

We see from Eq. (4.21) that emissivity is a number between 0 and 1. A blackbody source has the maximum emissivity $\varepsilon = 1$ for all wavelengths. Because $\varepsilon \leq 1$, the spectral exitance of any real source at a given temperature is bounded by the spectral exitance of a blackbody at the same temperature. Radiation from a thermal source (in equilibrium at temperature T) cannot exceed that of a blackbody at temperature T.

4.5.1 Kirchhoff's law

Consider radiation at a single wavelength λ incident on a flat surface. Three things can happen; the incident radiation can be reflected, transmitted, or absorbed, as shown in Eq. (4.22):

$$\phi_{\lambda,\,\mathrm{incident}}(\lambda) = \phi_{\lambda,\,\mathrm{reflected}}(\lambda) + \phi_{\lambda,\,\mathrm{transmitted}}(\lambda) + \phi_{\lambda,\,\mathrm{absorbed}}(\lambda). \tag{4.22}$$

Normalizing Eq. (4.22) to the incident spectral flux yields

$$1 = \frac{\phi_{\lambda,\,\mathrm{reflected}}(\lambda)}{\phi_{\lambda,\,\mathrm{incident}}(\lambda)} + \frac{\phi_{\lambda,\,\mathrm{transmitted}}(\lambda)}{\phi_{\lambda,\,\mathrm{incident}}(\lambda)} + \frac{\phi_{\lambda,\,\mathrm{absorbed}}(\lambda)}{\phi_{\lambda,\,\mathrm{incident}}(\lambda)} \equiv \rho(\lambda) + \tau(\lambda) + \alpha(\lambda). \tag{4.23}$$

Equation (4.23) defines spectral reflectance, transmittance, and absorptance at the wavelength of the incident radiation. If the surface is opaque, then $\tau(\lambda) = 0$ and $\alpha(\lambda) = 1 - \rho(\lambda) = \varepsilon(\lambda)$. Kirchhoff's law states that, for opaque materials, spectral absorptance equals spectral emissivity:

$$\alpha(\lambda) = \varepsilon(\lambda). \tag{4.24}$$

The equality of Eq. (4.24) implies that good reflectors (like polished metals) tend to be poor emitters and that good absorbers tend to be good emitters. Equation (4.24) does not mean that the same total power is absorbed and emitted by a receiver. Calculation of power emitted and absorbed requires an integration over wavelength, considering the spectral distribution of flux from both the source and the receiver which, from Eq. (4.3), depends on both T_{source} and T_{receiver}:

$$\phi_{\text{absorbed by receiver from source}} = A_{\text{receiver}} \times \int_0^\infty \alpha(\lambda) E_\lambda(\lambda, T_{\text{source}}) d\lambda \tag{4.25}$$

and

$$\phi_{\text{radiated by receiver}} = A_{\text{receiver}} \times \int_0^\infty \varepsilon(\lambda) M_\lambda(\lambda, T_{\text{receiver}}) d\lambda. \tag{4.26}$$

Obviously, $\alpha(\lambda) = \varepsilon(\lambda)$ does not make these two powers equal. The equality between spectral absorptance and spectral emissivity at a single wavelength does not imply equality of these quantities at other wavelengths. For instance, the spectral absorptance in the visible band for TiO_2 paint is around 0.2, while the spectral emissivity of the same paint in the LWIR is above 0.9 (Ref. 2).

4.5.2 Graybody

A graybody source has an emissivity that is less than unity and is independent of wavelength. The spectral exitance at any wavelength is a constant fraction ε of what a blackbody source at the same temperature would produce:

$$M_{\lambda,\text{graybody}}(\lambda, T) = \varepsilon M_{\lambda,\text{blackbody}}(\lambda, T). \tag{4.27}$$

This constant multiplier applies for integrated exitance as well, given that ε is flat with wavelength, implying that $M_{\text{graybody}} = \varepsilon\ \sigma\ T^4$. Graybody and blackbody curves have the same functional dependence on λ, as seen in Fig. 4.11, where we plot the spectral exitance as a function of wavelength for a series of 2000 K sources: a blackbody with $\varepsilon = 1.0$ and graybodies with $\varepsilon = 0.75$, 0.5, and 0.25.

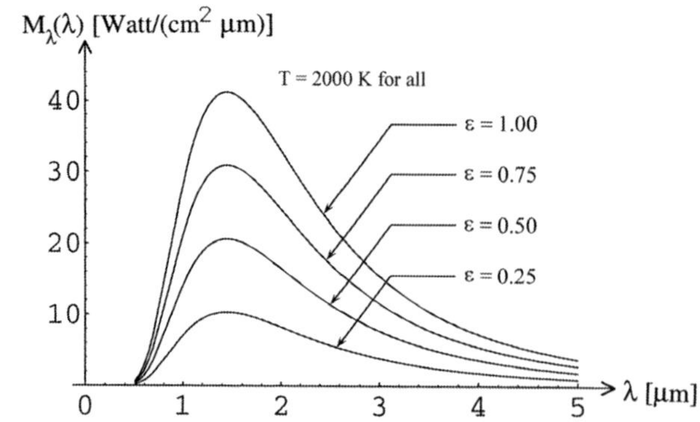

Figure 4.11 $M_\lambda(\lambda$, 2000 K) curves for $\varepsilon = 1.0$, 0.75, 0.5, and 0.25.

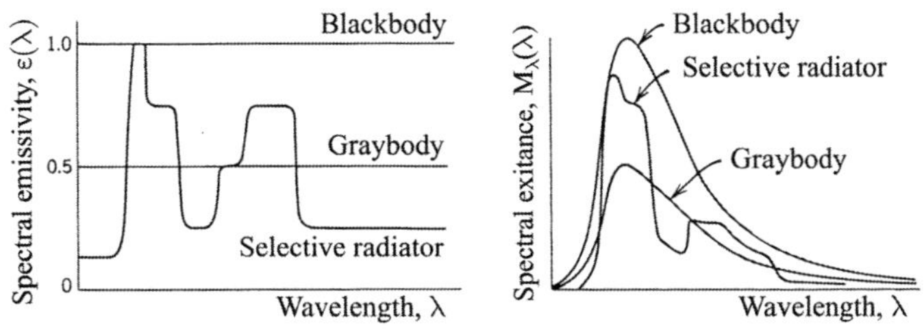

Figure 4.12 Comparison of plots of spectral emissivity and spectral exitance for a blackbody, a 50% emissivity graybody, and a selective radiator.

4.5.3 Selective radiator

A selective radiator is a source whose emissivity is not constant with wavelength. The spectral exitance at any λ of a selective radiator (at true temperature T) is the emissivity at that wavelength and temperature multiplied by Planck's blackbody function of Eq. (4.3) for that temperature:

$$M_{\lambda,\,\text{selective radiator}}(\lambda, T) = \varepsilon(\lambda, T) M_{\lambda,\,\text{blackbody}}(\lambda, T). \tag{4.28}$$

The plots in Fig. 4.12 show the relationships between spectral emissivity and spectral exitance for a blackbody, a graybody, and a selective radiator. For wavelengths where $\varepsilon = 1$, the spectral exitance of the selective radiator equals, but cannot exceed, that of a blackbody at the same temperature.

4.5.4 Emissivity of common materials

In general, ε is a function of both wavelength and temperature, but typically is a slow function of both. For engineering calculations, the usual specification for emissivity is just a single number for a given material unless more detailed information is available. Emissivity values for some common materials are seen in Table 4.2. The values quoted are average values for the long-wave infrared (LWIR) portion of the spectrum, for material temperatures around 300 K. Polished metal surfaces typically have a high reflectivity, a low absorption, and by Kirchhoff's law, Eq. (4.24), a corresponding low emissivity. Compared to a blackbody, they do not radiate much. Also, emissivity is strongly dependent on surface finish, oxidation, and surface contamination. A thin oil film, oxide coating, or surface roughness can increase emissivity by an order of magnitude compared to a polished metal surface.

4.5.5 Emissivity as a function of λ and *T*

Widely accessible data for emissivity of most materials tend to be like those of Table 4.2, that is, single numbers with no wavelength or temperature dependence. This approximation is useful for a first-order correction to the radiation emitted by an object. For metals in the infrared, emissivity tends to

Table 4.2 Room-temperature emissivity in the LWIR for common materials (from Ref. 2).

Emissivity of Various Common Materials			
Material	**Emissivity**	**Material**	**Emissivity**
Metals and their Oxides		**Other Materials**	
Aluminum: polished sheet	0.05	Brick: red common	0.93
sheet as received	0.09	Carbon: candle soot	0.95
anodized sheet, chromic acid process	0.55	graphite, field surface	0.98
vacuum deposited	0.04	Concrete	0.92
Brass: highly polished	0.03	Glass: polished plate	0.94
rubbed with 80-grit emery	0.20	Lacquer: white	0.92
oxidized	0.61	matte black	0.97
Copper: polished	0.05	Oil, lubricating (thin film on nickel base):	
heavily oxidized	0.78	nickel base alone	0.05
Gold: highly polished	0.02	oil film 1, 2, 5 $\times$ 10^{-3} in.	0.27, 0.46, 0.72
Iron: cast, polished	0.21	thick coating	0.82
cast, oxidized	0.64	Paint, oil: average of 16 colors	0.94
sheet, heavily rusted	0.69	Paper: white bond	0.93
Nickel: electroplated, polished	0.05	Plaster: rough coat	0.91
electroplated, no polish	0.11	Sand	0.90
oxidized	0.37	Skin, human	0.98
Silver: polished	0.03	Soil: dry	0.92
Stainless steel: type 18-8,	0.16	saturated with water	0.95
buffed type 18-8, oxidized	0.85	Water: distilled	0.96
Steel: polished	0.07	ice, smooth	0.96
oxidized	0.79	frost crystals	0.98
Tin: Commercial tin-plated	0.07	snow	0.85
sheet iron		Wood: planed oak	0.90

increase with temperature and decrease with wavelength. For any specific material or surface treatment, the dependence of emissivity on wavelength and temperature would need to be measured experimentally.

References

1. E. L. Dereniak and G. D. Boreman, *Infrared Detectors and Systems*, Wiley, New York, Chapter 2 (1996).
2. W. Wolfe and G. Zissis, *The Infrared Handbook, Revised Edition*, SPIE Press, Bellingham, Washington (1985).

Bibliography

Bramson, M. A., *Infrared Radiation: A Handbook for Applications*, Springer, New York (1968).

Taylor, J., *Radiation Exchange: An Introduction*, Academic Press, New York (1990).

Chapter 5
Detectors

Detectors are transducers, producing a measurable electrical output from collected radiation. The most important detector parameters are response speed, responsivity, and sensitivity. Regarding response speed, any signal of interest will vary with time. How fast can the detector follow changes in the input flux level? Responsivity, denoted by the symbol $\mathcal{R}$, is an output-per-unit-input quantity, e.g., [V/W]. We use it to predict the electrical output of the sensor given an input flux level. An important responsivity quantity is the spectral responsivity $\mathcal{R}(\lambda)$, the responsivity as a function of wavelength. Sensitivity is a separate quantity from responsivity. Sensitivity specifies the signal-to-noise ratio (SNR) that the user can expect for a given input flux level. The SNR is a crucial parameter in the determination of image detectability, that is, whether a given feature in the image can be reliably discerned in the presence of electronics noise.

5.1 Comparing Thermal Detectors and Photon Detectors

The two primary classes of infrared sensors are thermal detectors and photon detectors. Both respond to absorbed photons; however, their mechanism of response differs, leading to differences in response speed, spectral responsivity, and sensitivity. Thermal detectors (such as bolometers and pyroelectrics) absorb the energy of the photon as heat. This heat causes a temperature rise in the sensing element. The sensing element has a temperature-dependent electrical property, such as resistance. The change in this electrical property as a function of input flux level is measured by an external circuit. Thermal detectors are typically operated without significant cooling. On the other hand, photon detectors (such as photoconductors and photovoltaics) use the energy of the photon not as heat, but rather to increase the energy of a charge carrier so that the carrier makes an electronic transition across an energy gap. This transition is typically that of an electron from the valence band to the conduction band in a semiconductor material. The excitation of these carriers into a higher energy state affects the sensor's electrical properties. The change of electrical properties as a function of input flux level is measured by external circuitry. These sensors typically require cryogenic cooling for operation in the MWIR and LWIR bands. We begin by comparing the general properties of thermal and photon sensors in terms of response speed, sensitivity, and responsivity.

Thermal detectors tend to be slow because a finite time is required for the sensing element to rise in temperature after the absorption of energy. Typical time constants for thermal detectors are in the range of 1 to 100 msec. There is an inherent tradeoff between response speed and response magnitude for thermal detectors, as seen in Fig. 5.1. Because of their relatively long thermal time constants, waiting a longer time generally produces a larger response from a thermal detector.

Photon detectors are generally faster than thermal detectors because the electronic transition is virtually instantaneous upon photon absorption. Typical time constants for photon detectors are in the microsecond range or faster, determined by the resistance–capacitance (*RC*) product of the external readout circuit that interfaces with the photon detector or by the charge-carrier lifetime of the photogenerated carriers. To illustrate, consider a Ge photon detector, of cross-sectional area A and length L, shown in Fig. 5.2.

The typical resistivity of Ge is $\rho \sim 43$ ohm cm. The relative permittivity of Ge is $\varepsilon_r \sim 16$ ($\varepsilon_r = n^2$). For resistance and capacitance, we have $R = \rho L/A$ and $C = \varepsilon_0\varepsilon_r A/L$, respectively, so the time constant $\tau \equiv RC = \rho\varepsilon_0\varepsilon_r$ is independent of

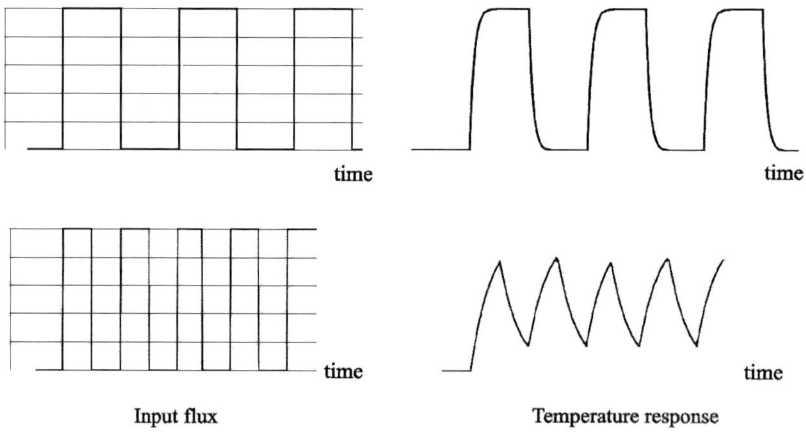

Figure 5.1 Tradeoff of response speed vs response magnitude for thermal detectors.

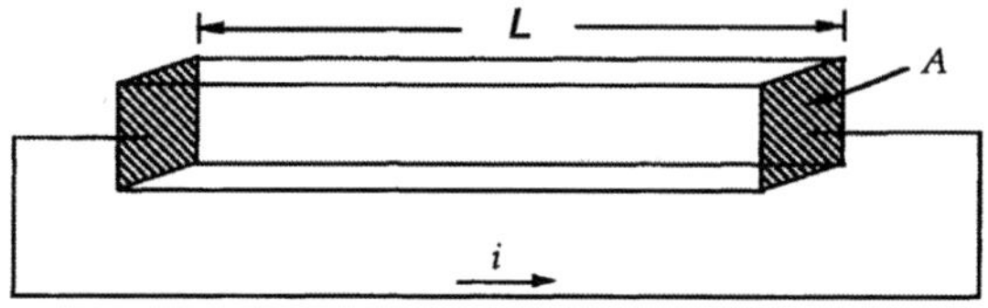

Figure 5.2 Estimating the response time of a photon sensor.

geometry. For these values, $\tau = \rho\varepsilon_0\varepsilon_r = 43$ ohm cm × 8.85×10^{-12} farad/m × 16, yielding $\tau_{sensor} \sim 6\times10^{-11}$ (V/A) (A sec/V) = 60 psec. In this case, however, the time response of the sensor is set by the transmission line required to extract the signal from the sensor. For a typical 1-ft co-axial cable, $R = 50$ ohm, $C = 30 \times 10^{-12}$ farad, yielding a time response of the cable as $\tau_{cable} \sim 1.5$ nsec. Integrated-circuit waveguides offer a somewhat faster response.

Photon detectors are generally more sensitive than thermal detectors by a factor of ~100 to 1000 because the fundamental noise limits are different for the two types of sensors. Photon-detector sensitivity is ultimately limited by the shot-noise fluctuations on the incoming photon stream, most often because of background photons. Thermal-detector sensitivity is ultimately limited by temperature fluctuations of the sensing element, considering its radiative connection with the ambient thermal environment. We will revisit these issues in more detail when we consider the fundamental noise mechanisms in Chapter 7.

5.2 Responsivity

As seen in Fig. 5.3, responsivity is the slope of the output-vs-input curve. We generally assume a constant slope of this curve for initial analyses.

We can describe responsivity in several ways, depending on the output and input variables. The output variables can be voltage, current, number of electrons, or a digital-output value. The input variables are typically either broadband watts from a blackbody, narrowband watts from a spectrally distributed source, or photons/sec, either narrowband or broadband. If we have narrowband flux that can be tuned to different wavelengths, we can characterize the spectral responsivity $\mathcal{R}(\lambda)$, which is the output of the detector per unit input, as a function of wavelength of incident flux. In situations where we must distinguish between current and voltage responsivity or between energy-based and photon-based quantities, subscripts are used to denote specifics, which include: $\mathcal{R}_{i,q}(\lambda)$ [A/(photon/sec)], $\mathcal{R}_{v,q}(\lambda)$ [V/(photon/sec)], $\mathcal{R}_i(\lambda)$ [A/W], and $\mathcal{R}_v(\lambda)$ [V/W]. Sometimes we will omit the subscript i or v if clear from the context which is being referred to, but we will always use the subscript q for photon-based quantities.

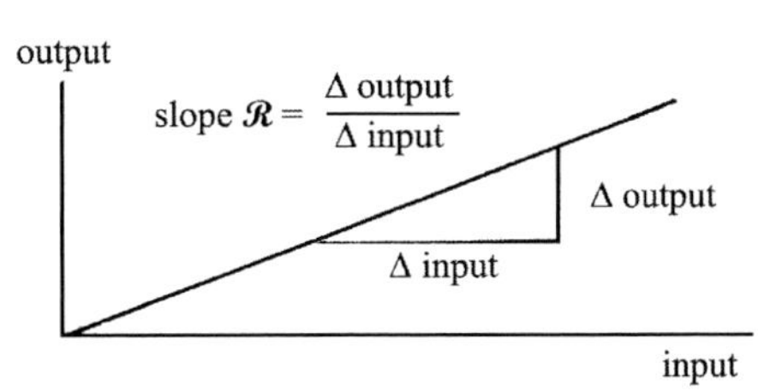

Figure 5.3 Responsivity is the ratio of output change to input change.

Another important responsivity quantity is $\eta(\lambda)$, the quantum efficiency [electrons per photon]. Quantum efficiency is defined statistically, assuming many incident photons, because for any given photon–electron interaction, either a photon is absorbed, or it isn't. For an ideal photon sensor, each absorbed photon causes the same response because an absorbed photon can take only one electron to the upper state. Quantum efficiency is the product of at least two terms, as seen in Eq. (5.1): the fraction of the incident radiation that passes the boundary from air into the material and the fraction of that radiation that is absorbed in the material:

$$\eta \approx (1 - r)(1 - e^{-a\ell}). \tag{5.1}$$

The Fresnel reflection r can be reduced by antireflection coatings. The thickness of the sensor material in the direction of photon propagation direction is ℓ, which is usually chosen so that nearly all the radiation passing the air/material boundary is absorbed inside the material. The absorption coefficient a [m^{-1}] is a strong function of wavelength, decreasing as the band edge is approached. Hence, $\eta(\lambda)$ falls as the cutoff wavelength is approached. For input flux described in photon-based units, the photogenerated current i [A] is

$$i\,[\mathrm{coul/sec}] = \eta\,[e^-/\mathrm{photon}]\,\phi_q[\mathrm{photon/sec}]\,q[\mathrm{coul}/e^-]. \tag{5.2}$$

In photon-based units [amp/(photon/sec)], the current responsivity is

$$\mathcal{R}_{i,q} = i/\phi_q = \eta\, q. \tag{5.3}$$

Note that photon-sensor responsivity is directly proportional to quantum efficiency. Thus, the spectral dependence of $\eta(\lambda)$ directly determines the spectral dependence of $\mathcal{R}(\lambda)$ in photon-based units. In energy-based units [A/W], the current responsivity of the photon sensor is

$$\mathcal{R}_i = \frac{i}{\phi} = \frac{\eta\,\phi_q\, q}{\phi_q(hc/\lambda)} = \frac{\lambda\eta\, q}{h\,c}. \tag{5.4}$$

Applying Eq. (5.4), we see that substituting $\eta = 0.10$ and $\lambda = 500$ nm produces the specified value (denoted by * on the graph in Fig. 5.4) of $\mathcal{R}_i = 40\ \mathrm{mA/W}$.

5.2.1 Spectral responsivity

The spectral responsivity of an ideal thermal detector is constant as a function of wavelength when expressed in energy-based units such as V/W, as seen in Fig. 5.5. One watt of radiation absorbed by the sensor at $\lambda = 1\ \mu\mathrm{m}$ will cause the same temperature change in the sensing element as 1 W of radiation at $\lambda = 10\ \mu\mathrm{m}$. No long-wavelength cutoff behavior is seen in thermal sensors because no energy gap is involved in the detection mechanism. Practically, the

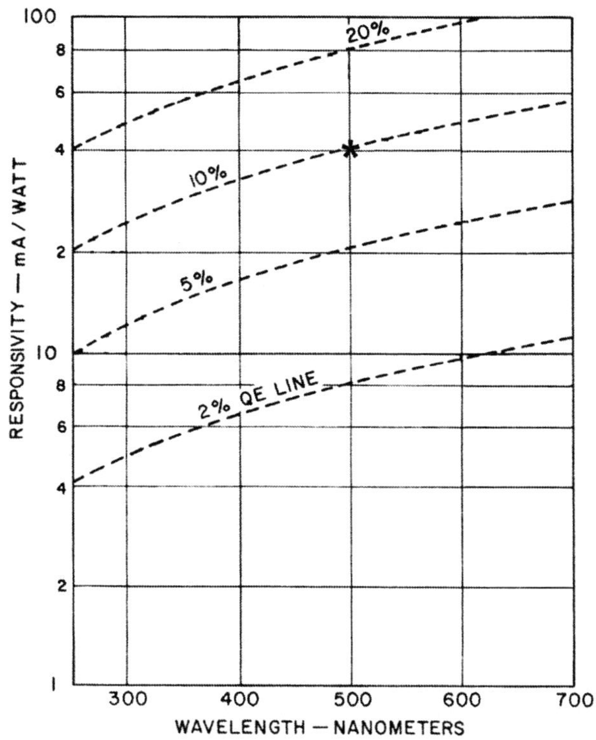

Figure 5.4 Plots of current responsivity vs wavelength for various values of quantum efficiency, for an ideal photon detector.

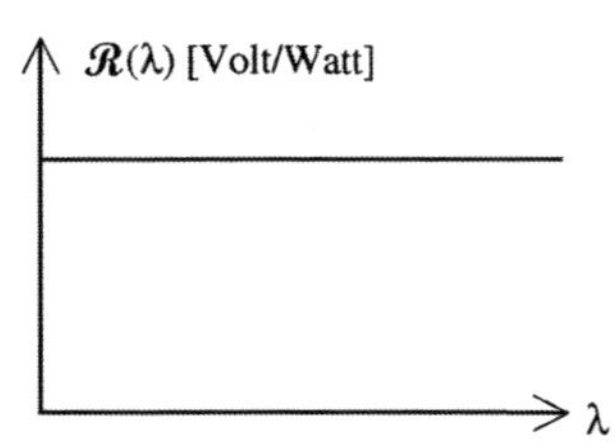

Figure 5.5 Ideal spectral responsivity for a thermal detector.

wavelength range for a thermal sensor is limited by the spectral absorption of the sensor material and the spectral transmission range of the window material that typically covers the sensor.

In practice, the non-ideal $\mathcal{R}(\lambda)$ for a thermal sensor will roll off at both very short and very long wavelengths, as seen in Fig. 5.6. At short wavelengths, the Fresnel reflectivity of the sensor will increase, consistent with an increased index of refraction, reducing the amount of radiation absorbed. At long wavelengths, the size of the sensor eventually becomes sub-wavelength, with reduced capture cross section for the radiation unless an antenna is used to couple radiation into the sensor. Spectral transmission of any window or coating in front of the sensor will also affect the spectral response.

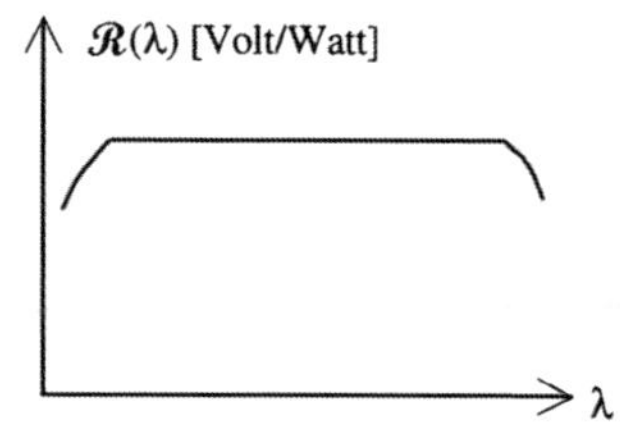

Figure 5.6 Non-ideal spectral responsivity for a thermal detector.

A photon detector has an inherent non-uniformity of response as a function of wavelength. For a photon incident on the sensor to be absorbed by the material and impart its energy to an electron, its energy must be sufficient to lift the electron across the energy gap of the sensor material. Recall the inverse relation of Eq. (3.3) between photon energy $\mathcal{E}$ and wavelength $\mathcal{E} = hc/\lambda$. Figure 5.7 shows a photon-sensor material with an energy gap $\mathcal{E}_{gap}$.

The wavelength at which the photon has just enough energy to raise the electron's energy state above $\mathcal{E}_{gap}$ corresponds to the long-wavelength cutoff λ_{cut}, which is the longest wavelength that will be detected by a sensor with a given $\mathcal{E}_{gap}$, as seen in Eq. (5.5):

$$\lambda_{cut} = hc/\mathcal{E}_{gap}. \tag{5.5}$$

Photons with wavelength longer than λ_{cut} are not absorbed and not detected. Spectral responsivity for an ideal photon detector goes to zero for photon wavelengths longer than λ_{cut}. Photons with additional energy beyond the required amount will place an electron into the continuum of conduction-band energy states, and thus these short-wavelength photons are detected. Because a photon can take at most one electron to an excited energy state, there is no additional response from any photon with energy exceeding $\mathcal{E}_{gap}$. We can solve for the long-wavelength cutoff of the photon detector in terms of $\mathcal{E}_{gap}$ in electron volts (eV):

$$\mathcal{E}_{gap}\,[\text{eV}] = \frac{hc}{\lambda_{cut}} = \frac{6.6\times10^{-34}\,\text{Js}\;3\times10^{8}\,\frac{\text{m}}{\text{s}}}{\lambda_{cut}\,[\mu\text{m}]\times10^{-6}\,\frac{\text{m}}{\mu\text{m}}} \times \frac{1\,\text{eV}}{1.6\times10^{-19}\,\text{J}} = \frac{1.24}{\lambda_{cut}\,[\mu\text{m}]}. \tag{5.6}$$

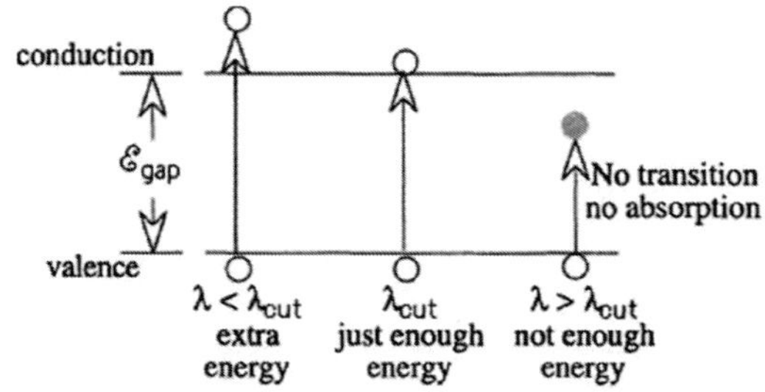

Figure 5.7 Origin of the cutoff wavelength for a photon detector.

If Si, with $\mathcal{E}_{gap} = 1.12$ eV, is used as a sensor, photons with wavelength shorter than 1.1 μm will be detected. If InSb with a smaller $\mathcal{E}_{gap} = 0.22$ eV is used, the sensor will detect photons with wavelength shorter than 5.6 μm.

Spectral responsivity curves $\mathcal{R}(\lambda)$ can be plotted with respect to either energy-derived or photon-derived units. Energy-derived units (e.g., V/W) are based on joules, while the other set of units is based on photons/sec. Photon detectors are most naturally described in photon-derived units but for historical reasons are often plotted with respect to energy-derived units. Because the proportionality between these two sets of units $\mathcal{E} = hc/\lambda$ depends on wavelength, these two sets of units will produce responsivity curves of different shapes for the same sensor, as seen in Fig. 5.8.

In photon-based units, the ideal photon detector has constant spectral responsivity as a function of wavelength, until the cutoff wavelength is reached, because any photon with sufficient energy will place a charge carrier in the conduction band. In energy-based units, the ideal photon-detector spectral responsivity is linearly proportional to wavelength:

$$\mathcal{R}(\lambda) = \frac{\lambda}{\lambda_{cut}} \mathcal{R}(\lambda_{cut}) \text{ when } \lambda \leq \lambda_{cut} \qquad \mathcal{R}(\lambda) = 0 \text{ when } \lambda > \lambda_{cut}. \tag{5.7}$$

We see a linear increase in spectral responsivity up to the cutoff wavelength because it takes more photons/sec to make a watt at longer wavelengths and because each photon with wavelength shorter that λ_{cut} produces the same electrical response from the sensor. For photons with $\lambda = 1$ μm, 1 W = 5×10^{18} photons/sec, and for photons with $\lambda = 10$ μm, 1 W = 50×10^{18} photons/sec. Because the sensor has equal electrical response to each photon with energy at or above $\mathcal{E}_{gap}$, it will appear to have a factor of 10 higher responsivity at 10 μm wavelength than at 1 μm wavelength if the spectral responsivity is plotted in energy-derived units.

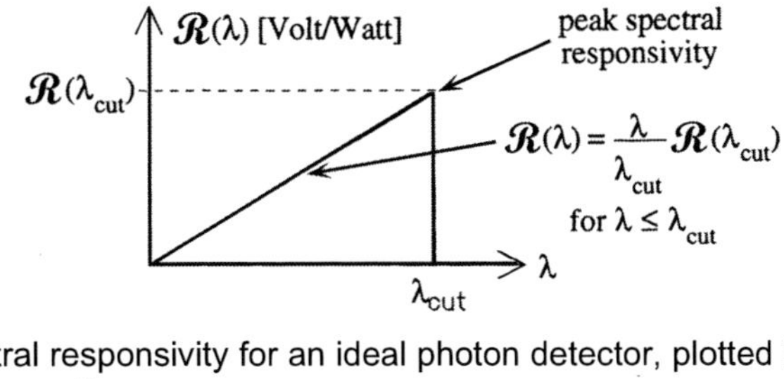

Figure 5.8 Spectral responsivity for an ideal photon detector, plotted in (top) photon units and (bottom) energy units.

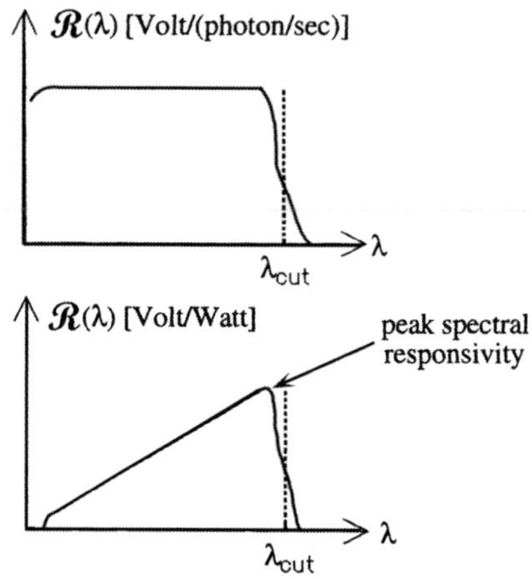

Figure 5.9 Spectral responsivity for a non-ideal photon detector, plotted in (top) photon units and (bottom) energy units.

In Fig. 5.9 we see that $\mathcal{R}(\lambda)$ for a non-ideal photon sensor rolls off at short wavelengths because the refractive index of the sensor material increases at short wavelengths, increasing the Fresnel reflection. At the long-wavelength end of the spectral response, the energy gap is not abrupt because the sensor material is at a finite temperature and the charge carriers consequently have some thermal energy. This reduces the peak response somewhat, and a residual response is seen for wavelengths slightly longer than λ_{cut}. When using manufacturers' specification sheets, note that $\mathcal{R}(\lambda)$ is often plotted on log–linear or log–log scales, which distorts the linear shape of the curves.

We use spectral responsivity to calculate the output from a detector in response to flux from a spectrally distributed source. Knowing the spectral responsivity of the sensor and the spectral flux falling on the detector $\phi_{\lambda,\mathrm{det}}(\lambda)$, we can calculate the sensor's electrical output using Eq. (5.8):

$$\text{Output} = \int_0^\infty \phi_{\lambda,\mathrm{det}}(\lambda)\ \mathcal{R}(\lambda)\mathrm{d}\lambda$$
$$[\mathrm{V}] = \left[\frac{\mathrm{W}}{\mu\mathrm{m}}\right]\left[\frac{\mathrm{V}}{\mathrm{W}}\right][\mu\mathrm{m}]. \tag{5.8}$$

We calculate the spectral flux falling on the detector from the source spectral exitance and the geometry of the source/sensor configuration as

$$\phi_{\lambda,\mathrm{det}}(\lambda) = \frac{M_\lambda(\lambda)}{\pi} A\Omega = L_\lambda A\Omega. \tag{5.9}$$

Using Eq. (5.8), we find that

$$\text{Output} = \frac{A\Omega}{\pi}\int_0^\infty M_\lambda(\lambda)\ \mathcal{R}(\lambda)d\lambda. \tag{5.10}$$

The factor $A\Omega/\pi$ [cm^2] is constant for a given source/sensor configuration. In the overlap integral of Eq. (5.8), there is a contribution to detector output only

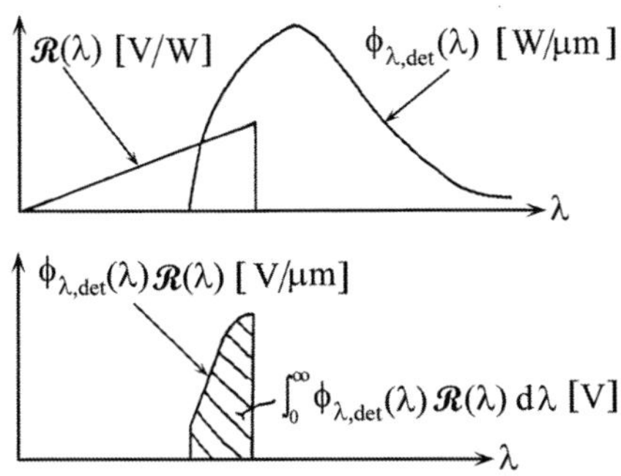

Figure 5.10 (top) Blackbody spectral flux and ideal photon detector spectral responsivity and (bottom) overlap integral of Eq. (5.8).

at those wavelengths where both $\phi_{\lambda,det}(\lambda)$ and $\mathcal{R}(\lambda)$ are nonzero, as seen in Fig. 5.10.

As an example of the calculation of the output of an ideal photon detector and a blackbody source, when both are specified in energy-based units, Eq. (5.8) becomes

$$\text{Output} = \frac{A\Omega}{\pi}\int_0^{\lambda_{cut}} \frac{2\pi hc^2}{\lambda^5(e^{hc/\lambda kT}-1)}\frac{\lambda}{\lambda_{cut}}\mathcal{R}(\lambda_{cut})\,d\lambda. \tag{5.11}$$

Because of the extra factor of λ, Eq. (5.11) is equivalent to an integration of the blackbody photon exitance [Fig. 4.7(a)] curve up to the cutoff wavelength using a constant $\mathcal{R}$, as seen in Fig. 5.11.

For a thermal detector having a constant spectral responsivity $\mathcal{R}$ and a blackbody source when both are specified in energy-based units as in Fig. 5.12, Eq. (5.8) becomes

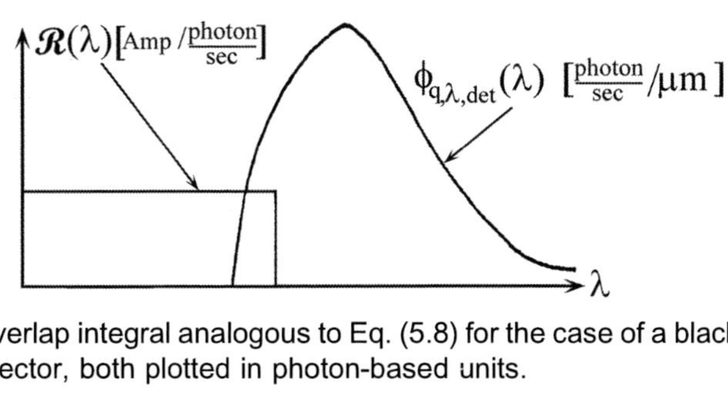

Figure 5.11 Overlap integral analogous to Eq. (5.8) for the case of a blackbody source and ideal photon detector, both plotted in photon-based units.

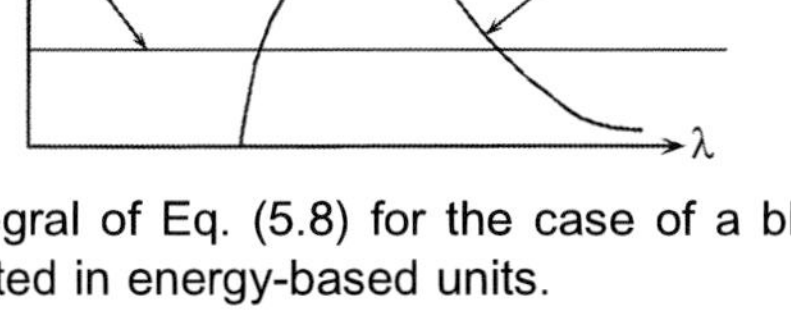

Figure 5.12 Overlap integral of Eq. (5.8) for the case of a blackbody source and ideal thermal detector, both plotted in energy-based units.

$$\text{Output} = \frac{A\Omega}{\pi}\mathcal{R}\int_0^\infty \frac{2\pi hc^2}{\lambda^5(e^{hc/\lambda kT}-1)}d\lambda = \frac{A\Omega}{\pi}\mathcal{R}\,\sigma T^4. \tag{5.12}$$

The response of a detector to a laser source can also be calculated using Eq. (5.8) by approximating the spectral flux from the laser as a delta function:

$$\phi_\lambda(\lambda) = \phi_{\text{laser}}\delta(\lambda - \lambda_{\text{laser}}). \tag{5.13}$$

Using the sifting property of the delta function, the detector output is then

$$\text{Output} = \int_0^\infty \phi_{\text{laser}}\delta(\lambda - \lambda_{\text{laser}})\,\mathcal{R}(\lambda)d\lambda = \phi_{\text{laser}}\,\mathcal{R}(\lambda_{\text{laser}}). \tag{5.14}$$

5.2.2 Blackbody responsivity

Alternatively, given a blackbody source at temperature T, we can measure the blackbody responsivity $\mathcal{R}(T)$, which is the output of the detector per watt of blackbody power falling on the sensor *at all wavelengths*. This is a function of the temperature of the blackbody source and is implicitly a function of the spectral responsivity $\mathcal{R}(\lambda)$ of the sensor. This common laboratory measurement is easy to set up, requiring only a blackbody source. With a measurement of $\mathcal{R}(T)$, we can calculate the spectral responsivity under some idealizing assumptions.

We measure the output [V] from the detector, according to Eq. (5.8): $\text{Output} = \int_0^\infty \phi_{\lambda,\text{det}}(\lambda)\mathcal{R}(\lambda)d\lambda$, then divide by the total power on the detector at all wavelengths: $\phi = (\sigma T^4/\pi)A\Omega$. The blackbody responsivity $\mathcal{R}(T)$ [V/W] can be written as

$$\mathcal{R}(T) = \frac{\int_0^\infty \phi_{\lambda,\text{det}}(\lambda, T)\,\mathcal{R}(\lambda)\,d\lambda}{\dfrac{\sigma T^4}{\pi}A\Omega}. \tag{5.15}$$

The flux on the detector at all wavelengths comes into the denominator of Eq. (5.15) regardless of whether the detector is sensitive to all the blackbody radiation. The amount of output produced and hence $\mathcal{R}(T)$ is determined by the overlap in wavelength between the spectral responsivity of the detector and the spectral flux curve of the blackbody source at temperature T used in the measurement. Thus, the spectral response of the detector is implicit in the calculation of blackbody responsivity. If the source spectrum is well matched to the detector's response, $\mathcal{R}(T)$ will be high.

5.2.3 Relating $\mathcal{R}(T)$ to $\mathcal{R}(\lambda_{\text{cut}})$

Blackbody responsivity is more convenient to measure than spectral responsivity because the measurement does not require a spectrally tunable source such as a spectrometer. Spectrally tunable sources are difficult to calibrate radiometrically, while the radiation from a blackbody source is known once its temperature is set. However, from a design viewpoint, we are more interested in having $\mathcal{R}(\lambda)$ data.

The typical procedure for IR photon-detector measurements is to measure the response to a blackbody at 500 K, and then calculate $\mathcal{R}(\lambda_{\text{cut}})$. The ideal photon-detector spectral responsivity of Eq. (5.7) is used: $\mathcal{R}(\lambda) = \lambda\mathcal{R}(\lambda_{\text{cut}})/\lambda_{\text{cut}}$ when $\lambda \leq \lambda_{\text{cut}}$ and $\mathcal{R}(\lambda) = 0$ when $\lambda > \lambda_{\text{cut}}$. Combining this responsivity with Eq. (5.15), we have

$$\mathcal{R}(T) = \frac{\int_0^{\lambda_{\text{cut}}} \frac{M_\lambda(\lambda, T)}{\pi} A\Omega \frac{\lambda}{\lambda_{\text{cut}}} \mathcal{R}(\lambda_{\text{cut}})\, d\lambda}{\frac{\sigma T^4}{\pi} A\Omega}. \tag{5.16}$$

Canceling common factors and bringing constants outside of the integral gives

$$\mathcal{R}(T) = \frac{\frac{\mathcal{R}(\lambda_{\text{cut}})}{\lambda_{\text{cut}}} \int_0^{\lambda_{\text{cut}}} M_\lambda(\lambda, T)\, \lambda\, d\lambda}{\sigma T^4}. \tag{5.17}$$

The relationship between $\mathcal{R}(T)$ and $\mathcal{R}(\lambda_{\text{cut}})$ defines the *K-factor* $K(\lambda_{\text{cut}}, T)$:

$$K(\lambda_{\text{cut}}, T) \equiv \frac{\mathcal{R}(\lambda_{\text{cut}})}{\mathcal{R}(T)} = \frac{\lambda_{\text{cut}} \sigma T^4}{\int_0^{\lambda_{\text{cut}}} M_\lambda(\lambda, T)\, \lambda\, d\lambda}. \tag{5.18}$$

A measurement of $\mathcal{R}(T)$ can yield a value of $\mathcal{R}(\lambda_{\text{cut}})$ given the cutoff wavelength of the sensor. Under the assumption of ideal spectral responsivity, this gives the photon-sensor spectral response $\mathcal{R}(\lambda)$ by means of Eq. (5.7). In Fig. 5.13, the K-factor is plotted as a function of cutoff wavelength, assuming a blackbody source at 500 K, the most common source temperature used in the characterization of LWIR sensors. A different source temperature would give a different K-factor curve. The K-factor is a number greater than 1, which is consistent with the observation that $\mathcal{R}(\lambda_{\text{cut}})$ is the maximum value of spectral responsivity, while $\mathcal{R}(T)$ is an average over many wavelengths. The minimum value of the K-factor curve indicates the best overlap between $\mathcal{R}(\lambda)$ and the spectral exitance of the measurement source, which yields the largest $\mathcal{R}(T)$ for a given sensor.

For example, a detector with a cutoff wavelength of 5 μm is measured to have $\mathcal{R}(T = 500\text{ K}) = 0.5$ V/W. From Fig. 5.13, we find that $K \sim 7.5$ for $\lambda_{\text{cut}} = 5$ μm. Thus, $\mathcal{R}(5\ \mu\text{m}) = 3.75$ V/W. Figure 5.14 shows the spectral response of the sensor obtained from this measurement under the ideal-response assumption of Eq. (5.7).

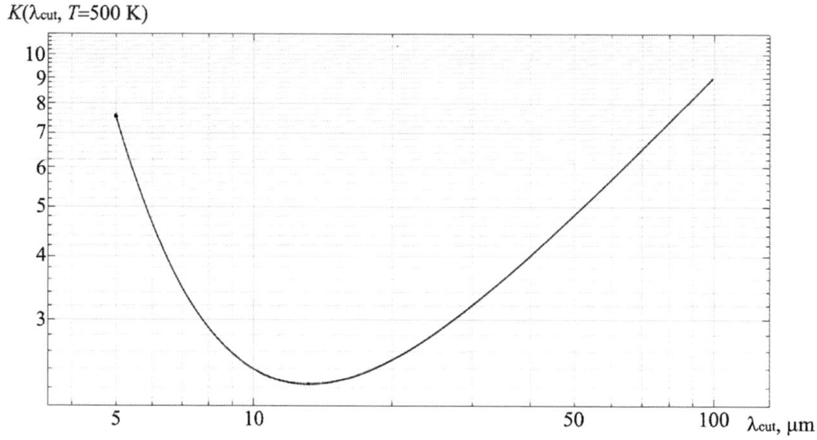

Figure 5.13 Ratio of peak spectral responsivity to blackbody responsivity as a function of cutoff wavelength, with the assumption of a 500 K blackbody source.

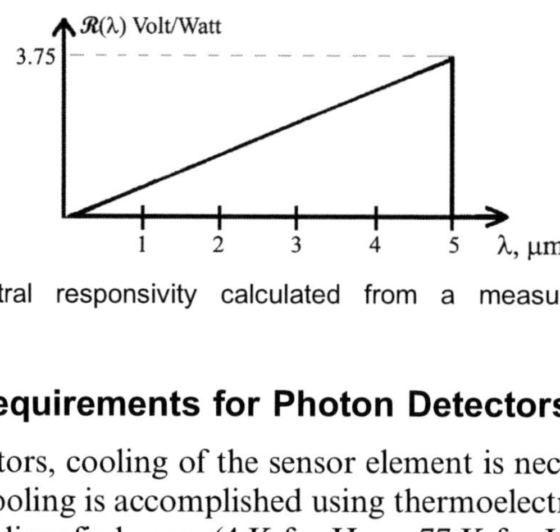

Figure 5.14 Spectral responsivity calculated from a measurement of blackbody responsivity.

5.3 Cooling Requirements for Photon Detectors

For photon detectors, cooling of the sensor element is necessary for detection of IR photons. Cooling is accomplished using thermoelectric coolers (down to 180 K), cryogenic liquefied gases (4 K for He or 77 K for LN2), or mechanical refrigerators (down to 60 K). Photon sensors operating in the 8- to 12-μm band are typically cooled to 77 K, and those operating in the 3- to 5-μm band are typically cooled to 180 K. Visible-band photon sensors are not typically cooled except in situations requiring long integration times such as astronomical observations. These cooling requirements can be understood in terms of the relationship between the bandgap energy and the cutoff wavelength. The charge carriers in a semiconductor have a thermal energy proportional to kT, where $k = 8.6 \times 10^{-5}$ eV/K. This gives a statistical range of energy levels for the charge carrier, as seen in Fig. 5.15. A higher temperature gives a larger spread of energy levels, allowing some electrons to thermally populate the conduction band.

If the detector material is not kept sufficiently cold, substantial thermal excitation of charge carriers across the energy gap of the semiconductor will

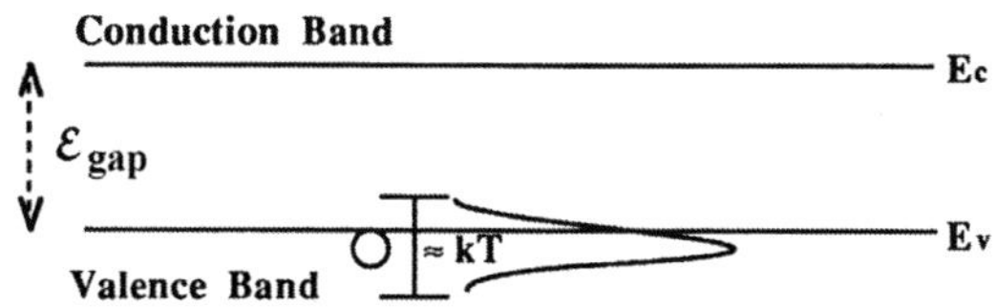

Figure 5.15 Sensor operating temperature determines the rate of thermal excitation of carriers across the bandgap.

occur. The problem is that there is shot noise (Section 7.6) on this thermally generated *dark current*. The shot noise makes detection of the photo-generated electrons more difficult. At room temperature (300 K), kT is an appreciable fraction of the bandgap for IR detectors. Long cutoff wavelengths require smaller energy gaps, hence, the sensor must be colder to avoid thermal excitation of charge carriers into the conduction band. Operation at a temperature such that kT is a few percent of $\mathcal{E}_{gap}$ gives generally satisfactory operation for integration times consistent with normal video rates of 30 to 60 frames/sec.

For example, using Si as a photon-detector material, its $\mathcal{E}_{gap} = 1.12$ eV leads to a cutoff wavelength of 1.1 μm. Typically, Si photodetectors are operated at a temperature of 300 K. In this case, $kT = 0.026$ eV, which is 2.3% of $\mathcal{E}_{gap}$. If InSb with $\mathcal{E}_{gap} = 0.22$ eV is used, the cutoff wavelength is 5.6 μm. An InSb photodetector is typically cooled to 180 K, so $kT = 0.015$ eV, which is 6.8% of $\mathcal{E}_{gap}$. Finally, a HgCdTe photodetector with a cutoff wavelength of 10 μm has $\mathcal{E}_{gap} = 0.12$ eV. Typical cooling for this sensor is 77 K, so $kT = 6.6$ meV, which is 5.5% of $\mathcal{E}_{gap}$. It is worth noting that thermal detectors are sometimes cryogenically cooled but not because of bandgap issues. We will return to cooling requirements for thermal detectors when we consider the fundamental noise mechanism for thermal detectors in Section 7.12.

5.4 Time Response and Frequency Response

Any measurement circuit acts as a low-pass filter. The noise-equivalent bandwidth Δf is the width of a flat-bandpass filter that passes the same amount of noise power as the circuit used to make the measurement. The noise-equivalent bandwidth is usually approximated as

$$\Delta f \approx \frac{1}{2\tau}, \tag{5.19}$$

where τ is the response time of the sensor. The exact equality depends on the functional form of the impulse response, such as exponential or square pulse, but a fast response always implies a broad bandwidth. Conversely, it takes more time to make a measurement with a narrow bandwidth (for instance, using a lock-in amplifier) compared to making a measurement with a wide bandwidth.

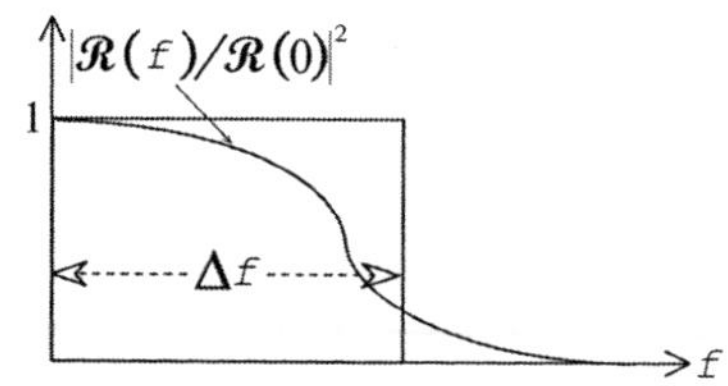

Figure 5.16 Noise-equivalent bandwidth.

From the Fourier transform of the temporal impulse response, we obtain $\mathcal{R}(f)$, the voltage transfer function. The absolute value squared gives the frequency response for power. When normalized to 1 at dc, this can be integrated to find the noise-equivalent bandwidth Δf, as seen in Eq. (5.20) and Fig. 5.16:

$$\Delta f = \int_0^\infty \left| \frac{\mathcal{R}(f)}{\mathcal{R}(f=0)} \right|^2 \mathrm{d}f. \tag{5.20}$$

Two functional forms of the temporal impulse response are commonly considered. The exponential has time response $v(t) = v_0\, e^{-t/\tau}$ step(t), and the averager has time response $v(t) = v_0$ rect[$(t-t_0)/\tau$]. The exponential impulse response arises from the measurement circuit's RC time constant, or from the finite charge-carrier lifetime. By taking the Fourier transform, the normalized magnitude of the voltage transfer function is seen in Eq. (5.21) and Fig. 5.17:

$$\left| \frac{\mathcal{R}(f)}{\mathcal{R}(0)} \right| = \frac{1}{\sqrt{1 + (2\pi\tau f)^2}}. \tag{5.21}$$

Carrying out the integration of Eq. (5.20), we find the noise-equivalent bandwidth $\Delta f = 1/(4\tau)$ for the exponential impulse response.

The averager integrates the input signal over a duration τ with an impulse response that is a rectangle function of width τ. From a Fourier transform of $v(t) = v_0$ rect[$(t-t_0)/\tau$], the normalized voltage transfer-function magnitude is expressed as Eq. (5.22) and shown in Fig. 5.18:

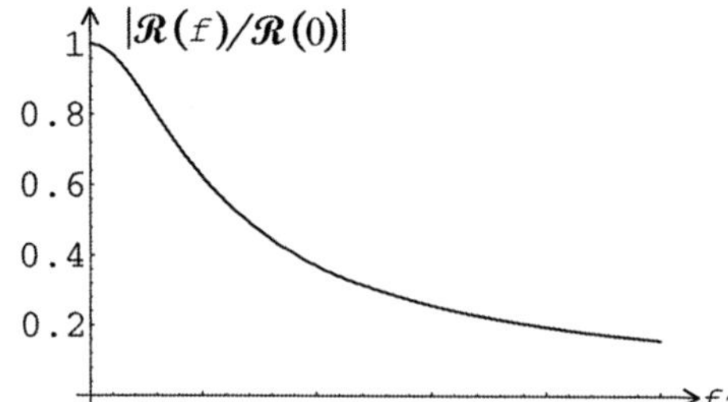

Figure 5.17 Normalized voltage-transfer function for an exponential impulse response.

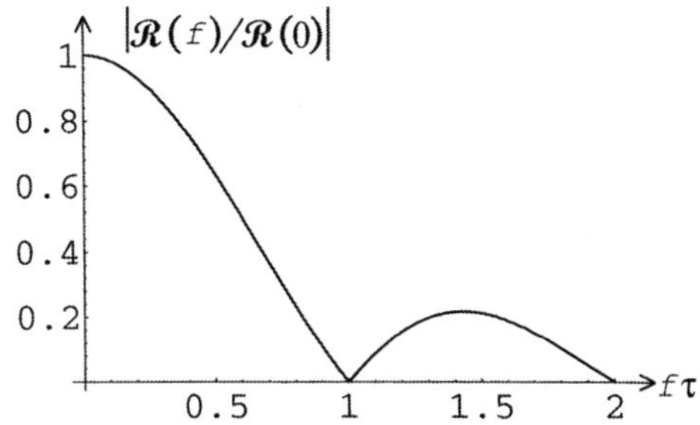

Figure 5.18 Normalized voltage-transfer function for an averager impulse response.

$$\left|\frac{\mathcal{R}(f)}{\mathcal{R}(0)}\right| = \left|\frac{\sin(\pi f \tau)}{\pi f \tau}\right|. \tag{5.22}$$

Carrying out the integration of Eq. (5.20), we find the noise-equivalent bandwidth as $\Delta f = 1/(2\tau)$ for the averager. This is the usual formula used to relate response time and noise-equivalent bandwidth.

5.5 Thermal Detectors

In a thermal detector, the absorption of a photon's energy causes a temperature rise in the sensing element. The sensing element has a temperature-dependent electrical property that is measured by an external circuit. Benefits of thermal detectors as compared to photon detectors are their broad spectral response and that no cooling is required. Drawbacks of thermal detectors are generally slower response and lower sensitivity.

5.5.1 Temporal response

We consider the heat-balance equation for thermal detectors with the following variable definitions. The difference between the detector temperature and the background temperature is $\Delta T = T_{\text{det}} - T_{\text{bkg}}$. The variable G_T is thermal conductance [watt/degree], which is characteristic of a particular thermal path. G_T is usually dominated by heat conduction through wires contacting the sensing element (allowing electrical readout). The minimum possible G_T is caused by radiation effects that arise because the sensor is radiatively coupled to its environment. The rate of heat gain or loss [watts] in the sensor is $\Delta\phi = G_T \times \Delta T$. The variable C_T is thermal capacitance [joule/degree] $= \Delta Q/\Delta T$, which is the energy absorbed per unit temperature rise. Thermal capacitance is proportional to the mass of the sensor element: $C_T = \text{mass} \times \text{specific heat}$ [joule/(gram degree)]. We can find specific heat values for various materials in handbooks or online. We then define a thermal time constant $\tau_T = C_T/G_T$. The differential equation for heat transfer in a thermal detector can be written as

$$C_T \frac{d(\Delta T)}{dt} + G_T \Delta T = \Delta\phi. \tag{5.23}$$

The sinusoidal steady-state solution to Eq. (5.23) can be obtained by substituting $\Delta T = \Delta T_0\, e^{j\omega t}$ and $\Delta\phi = \Delta\phi_0\, e^{j\omega t}$, where $\omega = 2\pi f$ is the radian modulation frequency of the incident flux (not the optical carrier frequency) and ΔT_0 and $\Delta\phi_0$ are the amplitudes of the variations of the temperature difference and incident flux modulation, respectively. This procedure yields $\Delta T_0 = \Delta\phi_0/(G_T + j\omega C_T)$, with magnitude

$$|\Delta T_0| = \frac{\Delta\phi_0}{\sqrt{G_T^2 + \omega^2 C_T^2}} = \frac{\Delta\phi_0}{G_T\sqrt{1+\omega^2\left(\frac{C_T}{G_T}\right)^2}} = \frac{\Delta\phi_0}{G_T\sqrt{1+\omega^2\tau_T^2}}. \tag{5.24}$$

The 3-dB modulation frequency at which the temperature response ΔT_0 is half of its value at $\omega = 0$ is $\omega_{3\text{-dB}} = 1/\tau_T = G_T/C_T$. A fast response is usually desirable. However, making $\tau_T = C_T/G_T$ small (for a fast response) also makes the temperature response $|\Delta T_0|$ small because $|\Delta T_0| = \Delta\phi_0/G_T$ at low frequencies. Because of this relationship, there is always a tradeoff of magnitude of response vs response speed for thermal detectors. The behavior can be seen in the log–log plot of Fig. 5.19, showing the thermal sensor responsivity (proportional to $|\Delta T_0|$) as a function of frequency. Note that the 3-dB frequency for any sensor configuration is referenced to the maximum value of responsivity (reached at $\omega = 0$), not to a predetermined reference level. This normalization is consistent with the finite dynamic range in terms of voltage for any amplifier. The 3-dB frequency is defined as the ω at which $\mathcal{R}$ has fallen off by half from its maximum value at $\omega = 0$.

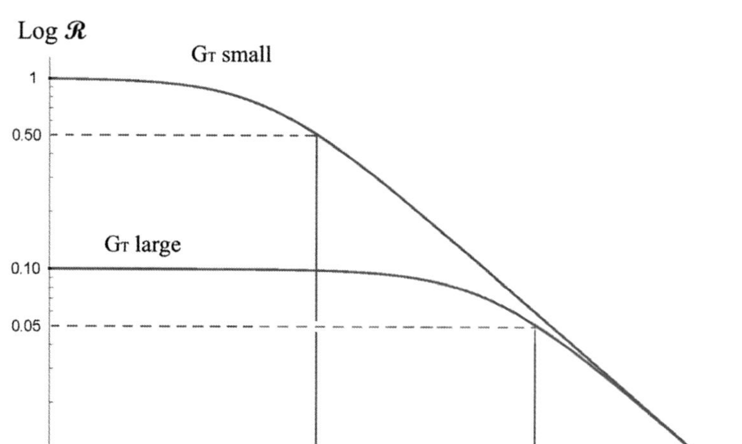

Figure 5.19 Log–log plot of thermal-sensor responsivity as a function of modulation frequency. Note that $\mathcal{R}$ is independent of G_T as $\omega \rightarrow \infty$.

At high modulation frequencies, the temperature response is independent of G_T. For a small thermal conductance, the sensor achieves a larger temperature response at low frequencies. However, the 3-dB frequency is lowered compared to the $\omega = 0$ value of $|\Delta T_0|$. A larger thermal conductance gives a faster response and a correspondingly higher 3-dB frequency. However, this faster frequency-response function has been normalized with respect to its correspondingly reduced low-frequency value. Thus, a fast response is obtained at the expense of a reduced response magnitude. The usual design for a thermal sensor is to keep both C_T and G_T small to achieve a fast response without sacrificing response magnitude. This leads to sensors with very small mass (minimizing C_T) and very thin wires (minimizing conductive contribution to G_T). The sensor may be placed in a vacuum enclosure to reduce the convective contribution to G_T.

The minimum-achievable value of G_T is from radiation effects. If we assume that the sensor is at a slightly elevated temperature with respect to the background, then $T_{det} = T_{bkg} + \Delta T$. The rate of heat loss is

$$\Delta\phi = \{(T + \Delta T)^4 - T^4\}\varepsilon\sigma A_{det} \tag{5.25}$$

$$\Delta\phi = \{[(T + \Delta T)(T + \Delta T)]^2 - T^4\}\varepsilon\sigma A_{det} \tag{5.26}$$

$$\begin{aligned}\Delta\phi &= \{[(T^2 + 2T\Delta T + \Delta T^2)]^2 - T^4\}\varepsilon\sigma A_{det}\\ &\approx \{[(T^2 + 2T\Delta T)]^2 - T^4\}\varepsilon\sigma A_{det}\end{aligned} \tag{5.27}$$

$$\Delta\phi \approx \{[T^4 + 4T^3\Delta T] - T^4\}\varepsilon\sigma A_{det} \tag{5.28}$$

$$\Delta\phi \approx 4T^3\Delta T\varepsilon\sigma A_{det}. \tag{5.29}$$

Thus, the minimum value of thermal conductance caused by radiation effects is

$$G_{T,\,radiation} \equiv \Delta\phi/\Delta T = 4T^3\varepsilon\sigma A_{det}. \tag{5.30}$$

5.5.2 Bolometric sensors

The bolometric sensor is a common type of thermal detector. A bolometer is a resistor with a temperature-dependent resistance. Typical bolometer resistances are in the range of 10 kΩ to 1 MΩ. Photons are captured on an absorber that is thermally contacted to the bolometer, and the energy of the absorbed photons cause a temperature rise in the sensor. The resulting change in resistance is sensed by an external circuit, as seen in Fig. 5.20.

A bias current is required to sense the change in resistance. The load resistor is often an element identical to the sensor but shielded from radiation. This arrangement allows the circuit to be insensitive to changes in ambient temperature. A figure of merit for bolometers is α, the fractional change in

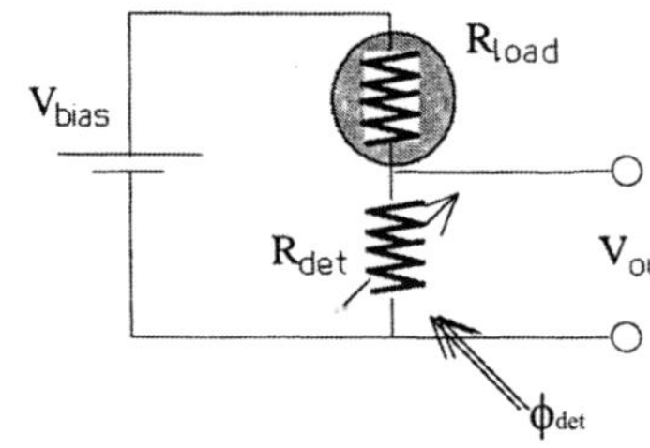

Figure 5.20 Bolometric readout circuit.

resistance per degree of temperature change, also called the temperature coefficient of resistance (TCR):

$$\alpha = \frac{1}{R}\frac{\partial R}{\partial T}. \tag{5.31}$$

Typically, α is positive (around 0.5% per °C) for metals because resistance of a metal rises with increased temperature. For superconducting materials held near the transition temperature, α is also positive and is orders of magnitude larger. For semiconductor materials, the resistance decreases with increasing temperature, giving a value of around −5% per °C for semiconductors at room temperature. The load resistor of Fig. 5.20 limits the current flow if the detector resistance decreases, avoiding the condition of *thermal runaway*.

The primary advantage of bolometers is their wide spectral response from the visible to the LWIR. Bolometers are useful in applications where cryogenic cooling is not feasible (from a weight, power, or cost viewpoint) but where IR detection is required at wavelengths that would require photon detectors to be cooled. Time constants are typically in the range of 1 to 100 msec, with minimum detectable powers in the range of 10^{-8} to 10^{-10} W. Better sensitivities are obtained for the longer time constants.

To develop an expression for bolometer responsivity, we begin by expressing the voltage across the detector:

$$V_{out} = iR_{det} = \frac{V_{bias}}{R_{load} + R_{det}} R_{det} = \frac{V_{bias}}{R_{load}/R_{det} + 1} \tag{5.32}$$

$$\frac{\partial V_{out}}{\partial R_{det}} = \frac{\partial}{\partial R_{det}} V_{bias}(R_{load}R_{det}^{-1} + 1)^{-1} = (-1)V_{bias}(R_{load}R_{det}^{-1} + 1)^{-2}(-1)\frac{R_{load}}{R_{det}^2} \tag{5.33}$$

$$\frac{\partial V_{out}}{\partial R_{det}} = \frac{V_{bias}}{\left(\dfrac{R_{load}}{R_{det}} + 1\right)^2}\frac{R_{load}}{R_{det}^2} = \frac{V_{bias}R_{load}}{(R_{load} + R_{det})^2} \tag{5.34}$$

$$\Delta V_{\text{out}} = \Delta R_{\text{det}} \frac{V_{\text{bias}} R_{\text{load}}}{(R_{\text{load}} + R_{\text{det}})^2}. \tag{5.35}$$

We can express the voltage responsivity as $\mathcal{R}_v \equiv \Delta V_{\text{out}}/\Delta\phi_0$. From Eq. (5.31), $\alpha = \frac{1}{R_{\text{det}}}\frac{\partial R_{\text{det}}}{\partial T}$, so $\Delta T_0 \approx \frac{\Delta R_{\text{det}}}{R_{\text{det}}\alpha}$. From our previous analysis of thermal detectors and Eq. (5.24), $|\Delta T_0| = \Delta\phi_0/(G_{\text{T}}\sqrt{1+\omega^2\tau_{\text{T}}^2})$. Substituting the expression for ΔT_0, we find that

$$\frac{\Delta\phi_0}{G_{\text{T}}\sqrt{1+\omega^2\tau_{\text{T}}^2}} = \frac{\Delta R_{\text{det}}}{R_{\text{det}}\alpha}. \tag{5.36}$$

From Eq. (5.35), we obtain

$$\Delta R_{\text{det}} = \Delta V_{\text{out}} \frac{(R_{\text{load}} + R_{\text{det}})^2}{V_{\text{bias}} R_{\text{load}}}. \tag{5.37}$$

Combining Eqs. (5.36) and (5.37), we find that

$$\frac{\Delta\phi_0}{G_{\text{T}}\sqrt{1+\omega^2\tau_{\text{T}}^2}} = \frac{\Delta V_{\text{out}} \frac{(R_{\text{load}}+R_{\text{det}})^2}{V_{\text{bias}} R_{\text{load}}}}{R_{\text{det}}\alpha}. \tag{5.38}$$

The final expression for voltage responsivity of a bolometer is

$$\mathcal{R}_v \equiv \Delta V_{\text{out}}/\Delta\phi_0 = \frac{V_{\text{bias}} R_{\text{load}} R_{\text{det}}\alpha}{(R_{\text{load}} + R_{\text{det}})^2 G_{\text{T}}\sqrt{1+\omega^2\tau_{\text{T}}^2}}. \tag{5.39}$$

5.5.3 Pyroelectric sensors

Another common thermal detector type is the pyroelectric sensor. Pyroelectric sensors are fabricated from materials (such as triglycine sulphate or lithium tantalate) of sufficient molecular asymmetry that they possess a permanent dipole moment even in the absence of an applied electric field. The magnitude of this dipole moment is temperature dependent. As photons are absorbed, the temperature of the element is changed, and there is motion of bound charge on the surface of the material, corresponding to the change in dipole moment. If we place this material between the plates of a capacitor, as seen in Fig. 5.21, the motion of bound charge induces a current flow in the circuit connected to the plates in response to dT/dt. The magnitude of the current flow is

$$i = \Theta A_{\text{cap}} \frac{dT}{dt}, \tag{5.40}$$

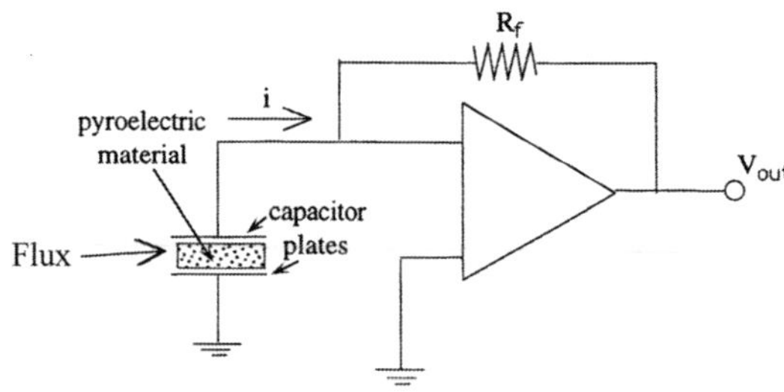

Figure 5.21 Pyroelectric sensor configuration.

where Θ is the pyroelectric coefficient (in the range of 10^{-8} coul cm^{-2} K^{-1}) and A_{cap} is the cross-sectional area of the capacitor. Equation (5.40) shows us that a pyroelectric detector responds only to a change in temperature.

The value chosen for the feedback resistance R_f affects both the responsivity and the response time. The output voltage is the product of the current of Eq. (5.40) and R_f, so a higher feedback resistance yields larger responsivity. However, the time response is affected by the circuit's RC time constant, which means that a higher feedback resistance gives a slower temporal response. Conversely, a small R_f gives a faster response but at the expense of a smaller responsivity. Typical values of R_f used with pyroelectric sensors are between 10^3 and 10^9 Ω.

For uncooled thermal detectors, pyroelectrics have good sensitivity with minimum detectable powers in the range of 10^{-9} to 10^{-10} W. The time response of pyroelectrics is fast for a thermal detector. Time constants of less than 100 nsec are possible, but with small responsivity. For many laser-measurement applications, this disadvantage is outweighed by the advantages of fast response, good sensitivity, and broad spectral response. Pyroelectric sensors are widely used in laboratory instrumentation and have also been fabricated as uncooled focal-plane arrays (FPAs). The required flux modulation for commercial pyroelectric-FPA-based imagers is generally accomplished by a rotating diffuser.

We will now develop an expression for current responsivity $\mathcal{R}_i$ for the pyroelectric sensor. We take $\Delta T = \Delta T_0\ e^{j\omega t}$, and from Eq. (5.24), $|\Delta T_0| = \Delta\phi_0/(G_T\sqrt{1+\omega^2\tau_T^2})$. We can substitute $\mathrm{d}\Delta T/\mathrm{d}t = \mathrm{d}T/\mathrm{d}t$ because the derivative is unchanged with a change in the reference temperature. We assume a transparent electrode, so $A_{det} = A_{cap}$. From Eq. (5.40), we find that

$$i = \Theta A_{cap}\frac{\mathrm{d}T}{\mathrm{d}t} = \Theta A_{det}\, j\omega \frac{\phi_0 e^{j\omega t}}{G_T\sqrt{1+\omega^2\tau_T^2}} \tag{5.41}$$

and

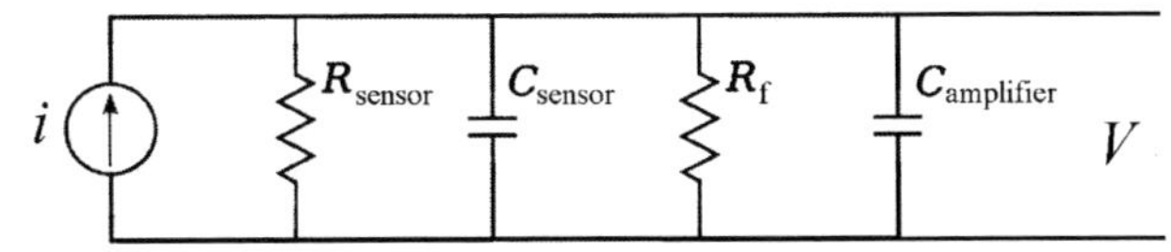

Figure 5.22 Equivalent circuit for a pyroelectric sensor.

$$\mathcal{R}_i = |i(t)|/\phi(t) = \frac{\Theta A_{det}\omega}{G_T\sqrt{1+\omega^2\tau_T^2}}. \tag{5.42}$$

This is the current responsivity of the pyroelectric sensor. It shows the effect of thermal time constant $\tau_T \equiv C_T/G_T$ and also shows that there is a decreased response at low modulation frequencies because of the linearity of Eq. (5.42) with ω. To develop the expression for voltage responsivity, we consider the equivalent circuit, as shown in Fig. 5.22. The usual setup is $C_{sensor} \gg C_{amplifier}$ and $R_{sensor} \gg R_f$, so we can approximate the circuit of Fig. 5.22, as seen in Fig. 5.23.

To obtain the voltage responsivity, we take the product of the current and the parallel impedance of the sensor and the feedback resistor. Defining an electrical time constant $\tau_E = R_f C_{sensor}$, we find that

$$V = \frac{iR_f}{\sqrt{1+\omega^2(R_f C_{sensor})^2}} = \frac{iR_f}{\sqrt{1+\omega^2\tau_E^2}}. \tag{5.43}$$

Combining this expression with Eq. (5.41),

$$V(t) = \frac{R_f}{\sqrt{1+\omega^2\tau_E^2}}\Theta A_{det} j\omega \frac{\phi_0 e^{j\omega t}}{G_T\sqrt{1+\omega^2\tau_T^2}}. \tag{5.44}$$

Using the definition for voltage responsivity, we find that

$$\mathcal{R}_v = \frac{|V(t)|}{\phi(t)} = \frac{R_f\Theta A_{det}\omega}{G_T\sqrt{1+\omega^2\tau_T^2}\sqrt{1+\omega^2\tau_E^2}}. \tag{5.45}$$

A pyroelectric sensor has both electrical and thermal time constants. The thermal time constant is typically longer. From Eq. (5.45), we see three

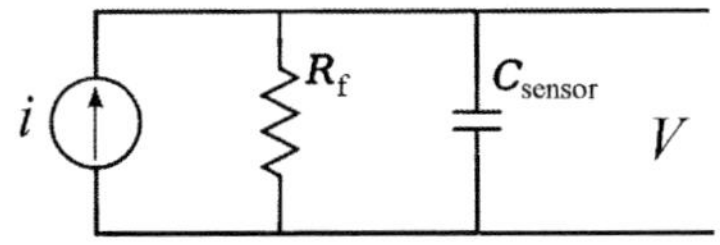

Figure 5.23 Approximate equivalent circuit for a pyroelectric sensor.

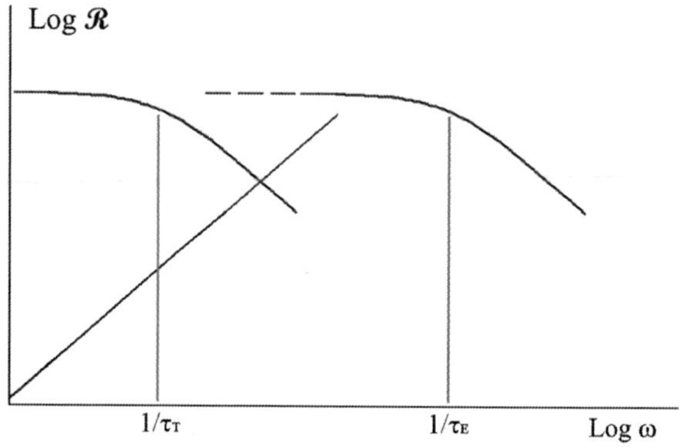

Figure 5.24 A pyroelectric sensor has three multiplicative frequency dependences.

multiplicative frequency-dependent contributions. The voltage responsivity $\mathcal{R}_v$ is usually plotted on a log–log scale, where multiplications are additive, as seen in Fig. 5.24.

First, $\mathcal{R}_v$ is linear with ω at low frequencies (no response at DC). Second, as ω increases past $1/\tau_T$ the thermal time constant takes over, so $\mathcal{R}_v$ is constant until $1/\tau_E$ is reached. Third, for frequencies higher than $1/\tau_E$, $\mathcal{R}_v$ rolls off because of the electrical time constant. Figure 5.25 illustrates the aggregate voltage responsivity for a pyroelectric sensor.

The value of the feedback resistor affects both $\mathcal{R}$ and τ_E, as seen in Fig. 5.26. A large feedback resistor gives a larger voltage response, but with a lower 3-dB frequency compared to the situation with a small feedback resistor. The thermal time constant τ_T is unchanged by the choice of R_f.

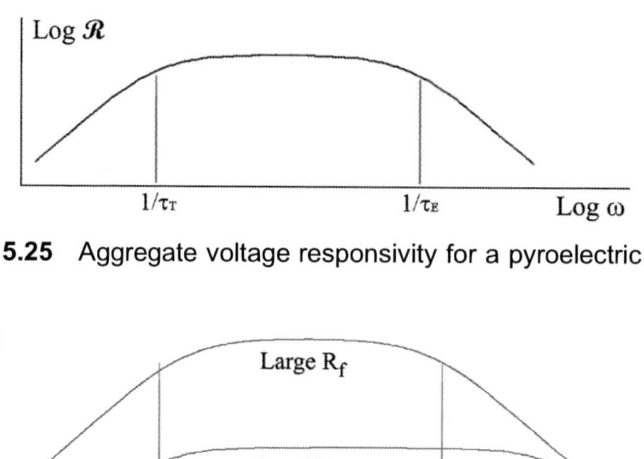

Figure 5.25 Aggregate voltage responsivity for a pyroelectric sensor.

Figure 5.26 Effect of feedback resistor on $\mathcal{R}$ and τ_E.

5.6 Photon Sensors

Photon detectors use the energy of the photon to increase the energy of a charge carrier so that the carrier makes a transition across an energy gap. The excitation of these carriers into a higher energy state affects the sensor's electrical properties. The change of electrical properties as a function of input flux level is measured by external circuitry. In general, as compared to thermal detectors, photon detectors have a faster response and higher sensitivity.

5.6.1 Photoconductive sensors

A photoconductive (PC) detector is constructed from homogeneous semi-conductor material, without a p–n junction. The mechanism for a PC detector is generation of electron–hole pairs by photon absorption in a semiconductor material. The material's conductivity is changed in response to detected photons. An external bias voltage and a voltage-divider circuit (similar to that used with a bolometer) are required to make this change in conductivity measurable as a resistance change. A PC detector cannot generate an open-circuit voltage. Minimum detectable powers for PC detectors are in the range of 10^{-10} to 10^{-12} W, with typical time constants in the microsecond-to-nanosecond range.

5.6.1.1 Intrinsic vs extrinsic PC

An intrinsic PC, seen in Fig. 5.27, uses natural semiconductor materials without appreciable doping. Electrons in the valence band are excited to the conduction band by a photon, leaving behind a hole in the valence band. Once the photon is absorbed, both electrons and holes are free to move. Electrons move in the conduction band and holes move in the valence band. However, usually the carrier mobilities are not equal, so typically one type of carrier dominates the current flow. We will use that simplifying assumption in our development of the current responsivity of the photoconductor in the following section.

An extrinsic PC, seen in Fig. 5.28, uses doped impurities (generally in Si or Ge host materials) to reduce $\mathcal{E}_{gap}$, facilitating detection at longer wavelengths. Typical dopants are Ar, Au, Cu, Hg, or In.

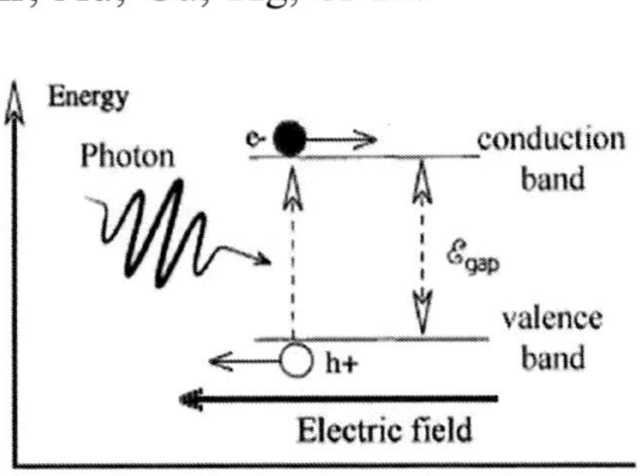

Figure 5.27 Intrinsic photoconductor.

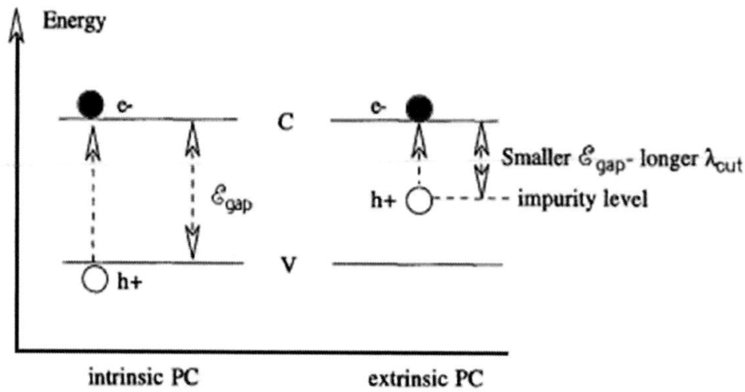

Figure 5.28 (left) Intrinsic vs (right) extrinsic photoconductor.

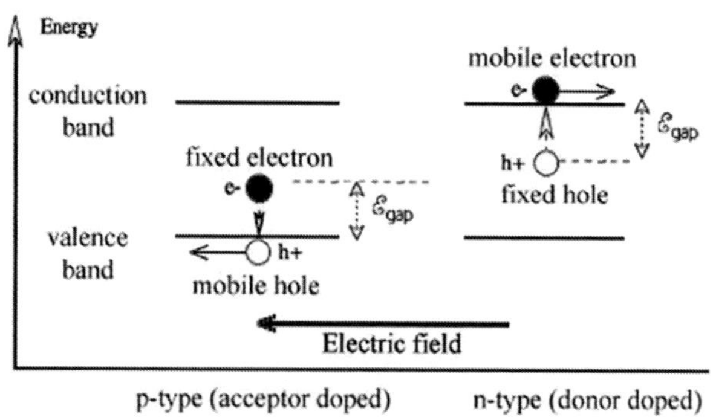

Figure 5.29 Extrinsic (left) p-type and (right) n-type photoconductors.

A donor-doped extrinsic (n-type) material has electrons doped into the energy gap. When excited to the conduction band by a photon, these electrons are mobile, leaving behind a fixed hole in the gap. An acceptor-doped extrinsic (p-type) material has holes doped into the energy gap. When excited to the valence band by a photon, these holes are mobile, leaving behind a fixed electron in the gap. Figure 5.29 compares these two situations.

Extrinsic PC materials have a lower absorption cross section for photons (implying a lower quantum efficiency η) because the doping concentrations achievable ($\sim 10^{15}/\text{cm}^3$) are much lower than the intrinsic atom concentrations ($\sim 10^{23}/\text{cm}^3$). Photons are absorbed according to Beer's law $\phi(z) = \phi_0 e^{-az}$, with the absorption coefficient a near the band edge for intrinsic PCs is $\sim 10^4$ cm^{-1} and for extrinsic PCs typical a is ~1–10 cm^{-1}. The spectral response of a PC sensor is determined by the material's $\mathcal{E}_{gap}$. For intrinsic materials, the cutoff wavelengths are seen in Table 5.1.

Two common extrinsics are Si:In with a cutoff of 7 μm, and Ge:Hg with a cutoff of 14 μm. Extrinsic PCs are used to get response to very long

Table 5.1 Cutoff wavelengths for intrinsic photoconductors.

Material	λ_{cut} [μm]
AlGaN	0.2–0.36*
CdS	0.52
ZnTe	0.54
CdSe	0.69
GaAs	0.80
CdTe	0.80
Si	1.1
GaAlAs	1.3
Ge	1.9
PbS	3.0
InAs	3.5
PbSe	4.5
InSb	5.6
HgCdTe	3–14*

*depends on the mixture of elements.

wavelengths in the far-IR/THz/mmW band (generally 20 to 100 μm in wavelength) that are not available from intrinsic materials. PC materials at these wavelengths have small bandgap energies, which require substantial cooling to minimize thermally generated dark current. For these PC sensors, liquid He (~ 4 K) is the cryogen of choice.

5.6.1.2 Photoconductive mechanism

In our development of an expression for current responsivity $\mathcal{R}_i$ for PC sensors, we will make the simplifying assumption that only electrons flow. A more complete treatment including contributions from both charge carriers can be found in Ref. 1. Relevant variables are electron concentration n [electrons/cm^3], electron mobility μ [cm^2/(V sec)], which is carrier velocity per unit **E** field, and conductivity $\sigma \approx n\mu q$ [1/(ohm cm)], where q is the charge on an electron. The general configuration of a PC sensor is shown in Fig. 5.30.

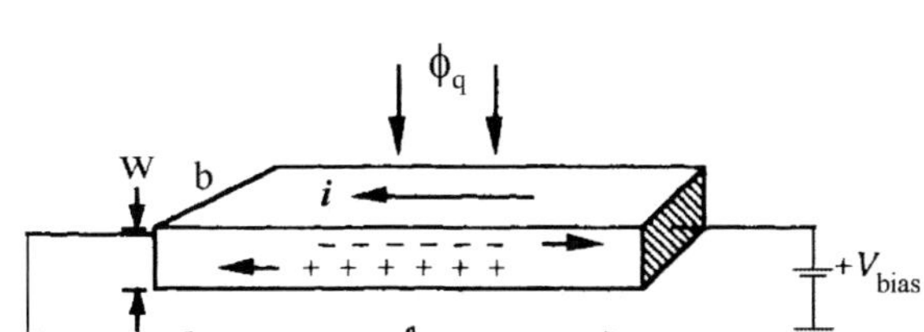

Figure 5.30 Configuration of a photoconductive sensor.

Once an electron is excited to the conduction band by photon absorption, it has a finite carrier lifetime τ in the upper state before it recombines with a hole. The photogenerated change in carrier concentration causes a change in conductivity, according to $\Delta\sigma = \Delta n \mu q$. We include both generation and recombination terms in the differential equation for Δn, seen in Eq. (5.46):

$$\frac{\mathrm{d}}{\mathrm{d}t}\Delta n = \frac{\eta\,\phi_q}{\ell wb} - \frac{\Delta n}{\tau}. \tag{5.46}$$

The steady-state solution is $\Delta n = \eta\phi_q\tau/(\ell wb)$. We can write an expression for the current flow as $i = V_{\mathrm{bias}}/R_{\mathrm{det}} = V_{\mathrm{bias}}wb\,\sigma/\ell$. This leads to an expression for the change in current for a change in conductivity $\Delta i/\Delta\sigma = V_{\mathrm{bias}}wb/\ell$, leading to the expression for (photon-based) current responsivity seen in Eq. (5.48):

$$\Delta i = \Delta\sigma\frac{V_{\mathrm{bias}}wb}{\ell} = \Delta n\frac{V_{\mathrm{bias}}wb}{\ell} = \frac{\eta\phi_q\tau}{\ell wb}\mu q\frac{V_{\mathrm{bias}}wb}{\ell} \tag{5.47}$$

$$\mathcal{R}_{q,i}\left[\frac{\mathrm{amp}}{\mathrm{photon/sec}}\right] = \frac{\Delta i}{\phi_q} = \frac{\eta\tau}{\ell wb}\mu q\frac{V_{\mathrm{bias}}wb}{\ell} = \frac{\eta\tau}{\ell}\mu q\frac{V_{\mathrm{bias}}}{\ell}. \tag{5.48}$$

The electric field in the PC material is $\mathbf{E} = V_{\mathrm{bias}}/\ell$. With this substitution in Eq. (5.48), we find that

$$\mathcal{R}_{q,i}\left[\frac{\mathrm{amp}}{\mathrm{photon/sec}}\right] = \eta q\frac{\tau\mu\mathbf{E}}{\ell}. \tag{5.49}$$

Comparing this to Eq. (5.3), the expression for $\mathcal{R}_{q,i} = \eta\,q$ for a generic photon detector, we see an extra factor called photo-conductive gain $G = \tau\mu\mathbf{E}/\ell$. The numerator $\tau\mu\mathbf{E}$ is the distance an electron travels before recombination. If the contacts at each end of the PC material are ohmic (no potential energy barrier), when an electron leaves one end, an electron enters at the other end to maintain charge neutrality. The entering electron retains the original generation time as the start of its lifetime.[1] Both generation and recombination are random processes. If the electron's lifetime before recombination τ is longer than the electron transit time through the device ($\ell/(\mu\mathbf{E})$, then the current flowing in response to ϕ_q is multiplied by the factor $G > 1$. The photoconductive gain G is the average number of electrons flowing in the circuit for each absorbed photon. The gain G can be controlled by the voltage applied across the device. Note that because $G \propto \tau$, a large gain and a fast response are mutually exclusive.

5.6.2 Photovoltaic mechanism

In a photovoltaic (PV) detector, absorbed photons generate hole–electron pairs at a p–n junction in a semiconductor. A built-in electric field exists at the p–n junction, which causes immediate separation of the hole–electron pair, as

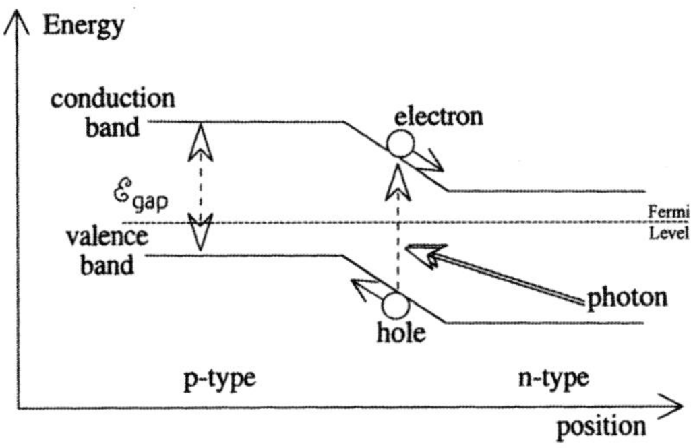

Figure 5.31 Energy-band diagram for a photovoltaic sensor.

seen in Fig. 5.31. This built-in field leads to the main operational difference between PC and PV detectors. A PV detector can generate a voltage across an open circuit, whereas a PC detector cannot. While a PV detector can operate in a mode where it is biased, an external bias voltage is not required for the sensor to operate. The cutoff wavelengths for PV detectors are the same as for PC detectors (Table 5.1), depending on the bandgap energy of the semiconductor. Extrinsic materials are not generally used in PV sensors because of the requirement to create a p–n junction, so the long-wavelength responses seen for extrinsic PC detectors are not achievable for PV detectors. The sensitivity of PV sensors is similar to that for PC sensors, with minimum detectable powers in the range of 10^{-10} to 10^{-12} W. The limiting sensitivity for a PV sensor is $\sqrt{2}$ better than that for a PC sensor because the recombination process in PV sensors is deterministic rather than random. Time constants for PV sensors are generally in the microsecond-to-nanosecond range, with the reverse-biased mode providing the fastest response ($\approx$ 100 psec).

The p–n junction in a PV detector is a diode, so the current–voltage (I–V) characteristic is that of a diode, as shown in Fig. 5.32, which also shows a diagram indicating the sign convention for positive current and positive voltage.

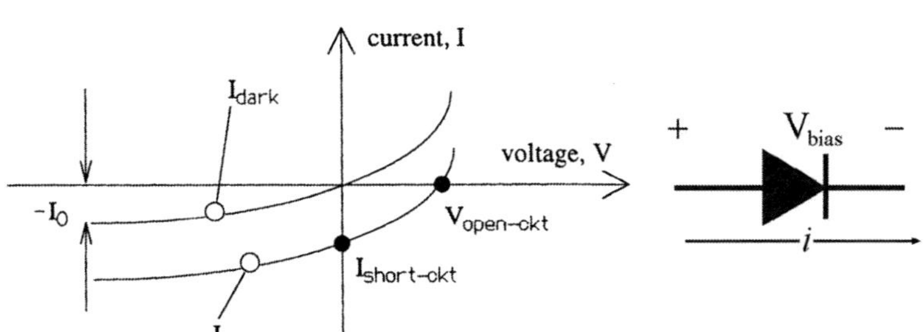

Figure 5.32 (left) Current-voltage curve for a photovoltaic sensor and (right) the sign convention for positive voltage and current.

With no flux incident on the sensor, the current–voltage relationship for a PV sensor is

$$I_{\text{dark}} = I_0(e^{qV/\beta kT} - 1), \tag{5.50}$$

where β is an empirical factor for the specific diode, q is the electronic charge, and k is Boltzmann's constant. With incident photon flux ϕ_q, the photo-generated current adds in

$$I_{\text{total}} = I_0(e^{qV/\beta kT} - 1) - \eta\,\phi_q q, \tag{5.51}$$

where the negative sign arises from the sign convention for positive current in a diode, as seen in Fig. 5.32, and η is the quantum efficiency [electrons/photon]. Three operating modes are possible for a PV detector: open-circuit, short-circuit, and reverse-bias. In each case, we begin with the I–V relation for the illuminated diode of Eq. (5.51).

5.6.2.1 Open-circuit mode

The condition where I_{total} is held at 0 in Eq. (5.51) gives an open-circuit voltage proportional to the natural log of the photon flux, providing a large (yet nonlinear) dynamic range:

$$V_{\text{open-ckt}} = \frac{\beta kT}{q} \ln\left(\frac{\eta\phi_q q}{I_0} + 1\right). \tag{5.52}$$

Connecting the output of the sensor to one input node of an operational amplifier, as seen in Fig. 5.33, allows a voltage to develop across the sensor, but does not allow current to flow.

5.6.2.2 Short-circuit mode

We can obtain the short-circuit condition shown in Fig. 5.34 by holding the voltage V across the diode at 0 in Eq. (5.51). In that case, the current flow is linearly proportional to photon flux:

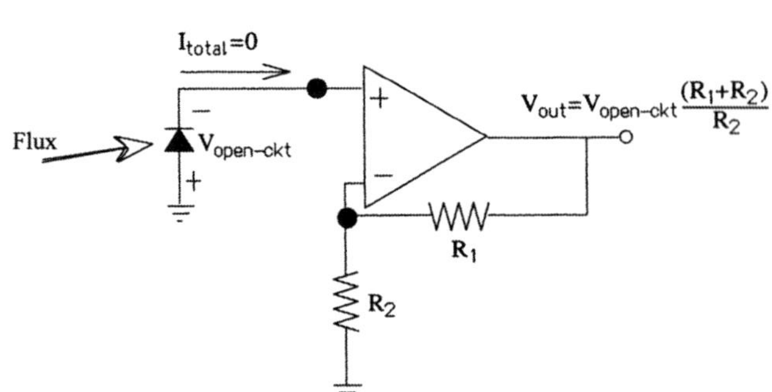

Figure 5.33 Open-circuit-mode amplifier configuration.

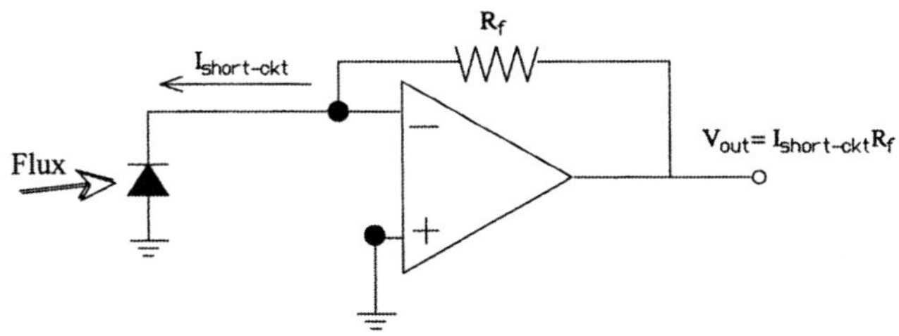

Figure 5.34 Short-circuit-mode amplifier configuration.

$$I_{\text{short-ckt}} = -\eta\, \phi_q q. \tag{5.53}$$

This situation is implemented by grounding one node of the operational amplifier (op-amp), which keeps the voltage equal to zero at the other node so that no voltage can develop when the sensor is illuminated. Thus, the photogenerated current flows through the feedback resistor, converting the current into an output voltage.

5.6.2.3 Reverse-bias mode

As seen in Fig. 5.35, the reverse-bias mode holds the diode voltage V of Eq. (5.51) at a negative value (polarity defined in Fig. 5.32) by connecting the bias voltage to one node of the op-amp. This configuration is also called the *photoconductive mode* of operation, given the similarities with PC operation. The current is linearly proportional to photon flux but with a reverse-saturation current I_0 added:

$$I_{\text{reverse-bias}} = -I_0 - \eta\, \phi_q q. \tag{5.54}$$

The advantage to this mode of operation is that the reverse bias increases the separation of the space-charge layer that constitutes the p–n junction, decreasing the diode capacitance, which has an inverse dependence on the separation distance. This lowers the RC time constant of the circuit, yielding a

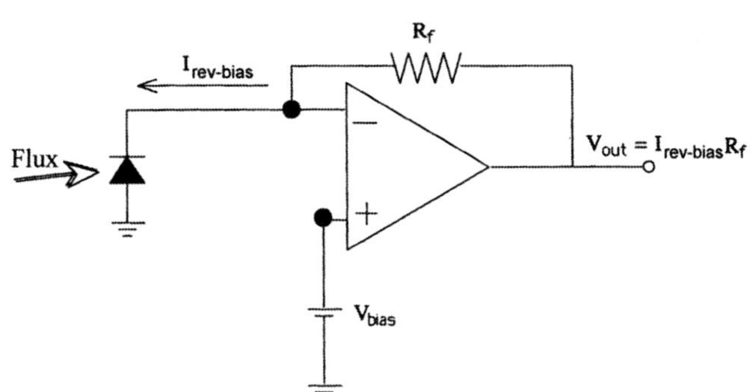

Figure 5.35 Reverse-bias-mode amplifier configuration.

faster response time. However, the reverse-saturation current will add an additional shot-noise term, somewhat decreasing the SNR.

5.6.3 Schottky-barrier detectors

Another photon detector is the Schottky barrier, seen in Fig. 5.36, which is fabricated of a thin (~10-nm) film of metal deposited onto a semiconductor substrate. Silicon is the most common substrate material because of the well-developed fabrication processes available for Si, although other semiconductors can be used. The metal–semiconductor junction is back-illuminated through the Si, with absorbed photons releasing electrons from the metal into the substrate, which are then collected as signal.

The quantum efficiency of the Schottky-barrier sensor's photo-generation process is an order of magnitude lower than that of PV materials such as Si and HgCdTe. However, because the quantum efficiency is independent of the exact thickness of the metal, Schottky-barrier detectors can be fabricated into FPAs that have excellent detector-to-detector response uniformity. This characteristic is valuable because spatial nonuniformity in IR FPAs is often the ultimate limit to image detectability, rather than temporal noise.

As seen in Fig. 5.37, the most common metals used for Si-based Schottky-barrier detectors are: Ir, with a cutoff wavelength of 9.5 μm; Pt, with a cutoff wavelength of 5.6 μm; Pd, with a cutoff wavelength of 2.6 μm; and Ni, with a

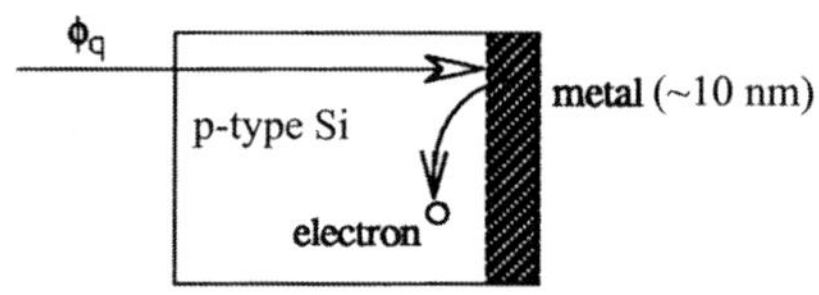

Figure 5.36 Schottky-barrier sensor mechanism.

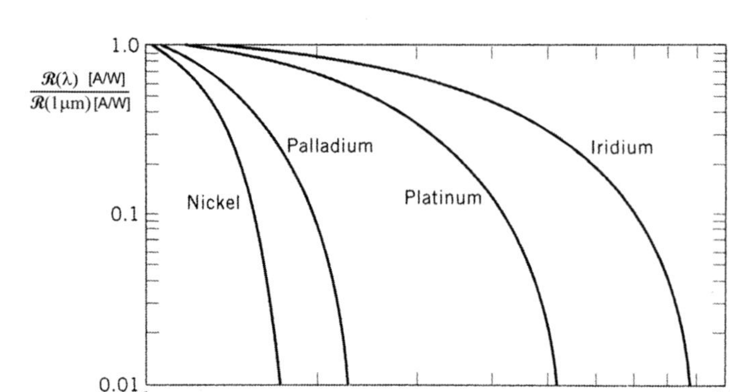

Figure 5.37 Responsivity for various metal-silicide Schottky-barrier detectors, normalized to responsivity at $\lambda = 1$ μm.

cutoff wavelength of 1.8 μm. Of these, PtSi is the most-developed technology, used for detector arrays in the 3- to 5-μm MWIR band.

Reference

1. R. H. Bube, *Photoconductivity of Solids*, Wiley, New York (1960).

Bibliography

Dereniak, E. L., and G. D. Boreman, *Infrared Detectors and Systems*, Wiley, New York (1996).

Dereniak, E. L., and D. G. Crowe, *Optical Radiation Detectors*, Wiley, New York (1984).

Keyes, R. J., Ed., *Optical and Infrared Detectors, 2nd Edition*, Springer, New York (1980).

Kinch, M. A., *Fundamentals of Infrared Detector Materials*, SPIE Press, Bellingham, Washington (2007) [doi: 10.1117/3.741688].

Kinch, M. A., *State-of-the-Art Infrared Detector Technology*, SPIE Press, Bellingham, Washington (2014) [doi: 10.1117/3.1002766].

Kruse, P. W., L. D. McGlauchlin, and R. McQuistan, *Elements of Infrared Technology*, Wiley, New York (1962).

Rieke, G. H., *Measuring the Universe: A Multiwavelength Perspective*, Cambridge (2017).

Rogalski, A., *Infrared Photon Detectors*, SPIE Press, Bellingham, Washington (1995).

Smith, R., F. Jones, and R. P. Chasmar, *The Detection and Measurement of Infrared Radiation, 2nd Edition*, Oxford University Press, Oxford (1968).

Vincent, J. D., S. E. Hodges, J. Vampola, M. Stegall, and G. Pierce, *Fundamentals of Infrared and Visible Detector Operation and Testing, 2nd Edition*, Wiley, New York (2016).

Chapter 6
Temperature Measurement

This chapter uses information from Chapters 4 and 5 to investigate how the temperature of a remote object can be estimated. If contact is possible, the true (kinetic) temperature of an object can be measured by a thermocouple. However, for a remote object that cannot be contacted, the only characteristic that can be measured is the radiation emitted from the object, which depends on both temperature and emissivity. Knowledge of spectral emissivity $\varepsilon(\lambda)$ of the object is necessary to avoid bias in the estimation of temperature. Measurement of wavelength-dependent emissivity requires specialized equipment, and often we must make assumptions about the source emissivity. A graybody approximation of a constant, non-unit emissivity can be made if more detailed information is not available.

One simple way we can estimate emissivity for a smooth opaque sample is to measure the specular reflection of a laser beam, then subtract the reflectance from unity to give absorption and hence emissivity. This measurement can be conveniently made with a laser, at its emission wavelength. However, measurement of diffuse reflectivity for rough surfaces requires more specialized apparatus.

As seen in Table 4.2, water has an LWIR emissivity of 96% because infrared radiation is strongly absorbed by water. This is particularly relevant to IR temperature measurement of biological samples, where we typically assume that the sample radiates as a blackbody with $\varepsilon = 1$, unless more detailed information is available.

6.1 Measurement Configurations

The optimum configuration for measurement of a remote object's temperature is one in which the region to be measured fills or overfills the FOV of the measuring instrument, as seen in Fig. 6.1. In this case, the flux received by the detector is the radiance of the source multiplied by area of the detector and the solid angle of the lens, as seen from the detector, $L_{source} \times A_{det} \times \Omega_{lens\ from\ img}$, per Eq. (3.23).

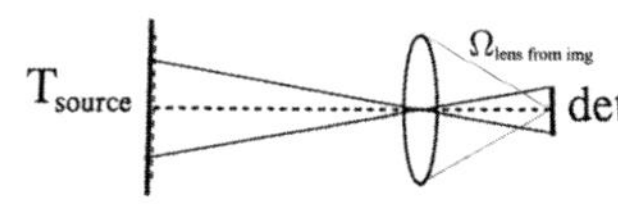

Figure 6.1 Configuration where the source for which temperature is to be measured overfills the FOV.

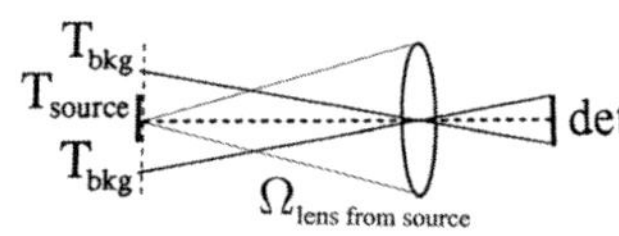

Figure 6.2 Configuration where the source for which temperature is to be measured does not fill the FOV.

If, as in Fig. 6.2, the source does not fill the FOV of the measurement instrument, the background will also contribute to the flux on the detector. In this case, it is best if this background is of significantly lower temperature than the region being measured. Under this assumption, the flux received by the detector is the radiance of the source multiplied by the area of the source and the solid angle of the lens as seen from the source, $L_{source} \times A_{source} \times \Omega_{lens\ from\ source}$, per Eq. (3.22).

As seen in Fig. 6.3, a simpler configuration can be considered, again under the assumption that the contribution from the background is negligible. In this case, the flux received by the detector is the radiance of the source multiplied by the $A\Omega$ product of the source/detector configuration, $\phi_{det} = L_{source} \times A_{source} \times \Omega_{det}$, per Eq. (3.11). This arrangement will typically collect less flux than the configuration of Fig. 6.2.

Three main methods exist for temperature measurement of a remote body: *radiation temperature* measures the source exitance integrated over all wavelengths; *brightness temperature* measures the source spectral exitance in a narrow band around one wavelength; and *color temperature* measures the source spectral exitance in narrow bands around two wavelengths, assessing the spectral composition of the source. These three measurements would ideally yield the true (kinetic) temperature of the source if the source is a blackbody.

We will consider the ideal cases of these three methods expressed in terms of measurement of source exitance, as the relevant $A\Omega$ products will be known for the specific measurement configuration used. We will then consider some practical issues involved in real temperature-measurement scenarios. In all cases, we will assume a Lambertian source in the calculations.

Figure 6.3 Configuration for temperature measurement without using a lens.

6.2 Radiation Temperature

The radiation temperature T_{rad} is the temperature of the blackbody that has the same integrated exitance as the object being measured:

$$M_{\text{measured}} = \int_0^\infty \varepsilon(\lambda) M_\lambda(\lambda, T_{\text{true}})\, d\lambda = \sigma T_{\text{rad}}^4. \tag{6.1}$$

If the object is a blackbody, then $T_{rad} = T_{true}$. If the object is a not a blackbody, then $T_{rad} < T_{true}$. If the object is a graybody with known constant emissivity ε, then T_{true} can be calculated from T_{rad} as follows:

$$M_{\text{measured}} = \varepsilon \int_0^\infty M_\lambda(\lambda, T_{\text{true}})\, d\lambda = \varepsilon\sigma T_{\text{true}}^4 = \sigma T_{\text{rad}}^4 \tag{6.2}$$

and

$$T_{\text{true}} = \frac{1}{\sqrt[4]{\varepsilon}} T_{\text{rad}}. \tag{6.3}$$

We illustrate this with the example of a graybody at 900 K with $\varepsilon = 0.8$, as shown in Fig. 6.4. Let us assume that, in the situation shown, 95 μW is measured as ϕ_{det}. The equation $\phi_{det} = (A\Omega/\pi)\sigma T_{rad}^4 = \sigma T_{rad}^4 A_s A_d/(\pi R^2) =$ 95 μW yields $T_{rad} = 851$ K. If the emissivity is known, the true temperature of 900 K can be calculated from this measurement using Eq. (6.3). If the emissivity is not known, the radiation temperature estimate is biased toward lower temperatures. Measurement of radiation temperature is similarly affected by any attenuation in the optics train, which can be an issue in environments where the optical surfaces can become dirty over time. For instance, in Fig. 6.4, if there is a window over the detector that transmits half of the incident radiation, then $\phi_{det} = 47.5$ μW, which yields T_{rad} = 715 K. Because of its simplicity, a measurement of radiation temperature is convenient, yet its estimate of the true temperature is not robust to incomplete knowledge of emissivity or attenuation.

6.3 Brightness Temperature

In the case of brightness temperature, the spectral exitance of the object is measured in a narrow wavelength region $\Delta\lambda$ around λ_0. The brightness temperature T_b is the temperature of the blackbody that has the same $M_{\lambda,\text{meas}}(\lambda_0)$ as the object being measured, as in Eq. (6.4):

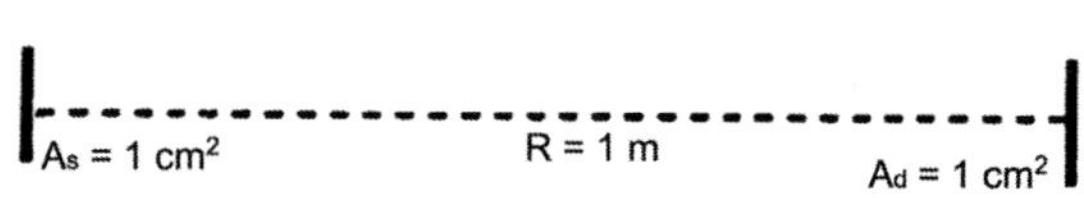

Figure 6.4 Example configuration for calculation of radiation temperature.

$$M_{\lambda,\,\text{meas}}(\lambda_0) = \frac{2\pi hc^2}{\lambda_0^5[e^{hc/\lambda_0 kT_b} - 1]}. \tag{6.4}$$

The brightness temperature is the temperature of the blackbody whose $M_\lambda(\lambda_0)$ equals the measured value, which is seen as the black dot in Fig. 6.5. For a particular point $M_\lambda(\lambda_0)$ on a plot of M_λ vs λ, only one blackbody curve passes through that point because a family of blackbody curves parameterized on temperature do not intersect one another, as seen in Fig. 4.6.

If an object has a measured $M_{\lambda,\text{meas}}(\lambda_0 = 3\ \mu\text{m}) = 0.39\ \text{W/(cm}^2\ \mu\text{m)}$, it has a brightness temperature of 800 K because the 800 K blackbody curve passes through this point on the plot. If we know the emissivity of the source at the measurement wavelength $\varepsilon(\lambda_0)$, then a brightness temperature measurement of $M_{\lambda,\text{meas}}(\lambda_0)$ can be corrected to give the true temperature as follows:

$$M_{\lambda,\,\text{meas}}(\lambda_0) = \varepsilon(\lambda_0)\frac{2\pi hc^2}{\lambda_0^5[e^{hc/\lambda_0 kT_{\text{true}}} - 1]}. \tag{6.5}$$

If the object emissivity at a 3-μm wavelength were known to be 0.51, then the measured $M_{\lambda,\text{meas}}(\lambda = 3\ \mu\text{m})$ of $0.39\ \text{W/(cm}^2\ \mu\text{m)}$ would be divided by the $\varepsilon(\lambda_0)$ of 0.51 to yield $0.76\ \text{W/(cm}^2\ \mu\text{m)}$, which corresponds to a true temperature of 900 K, as seen in Fig. 4.9.

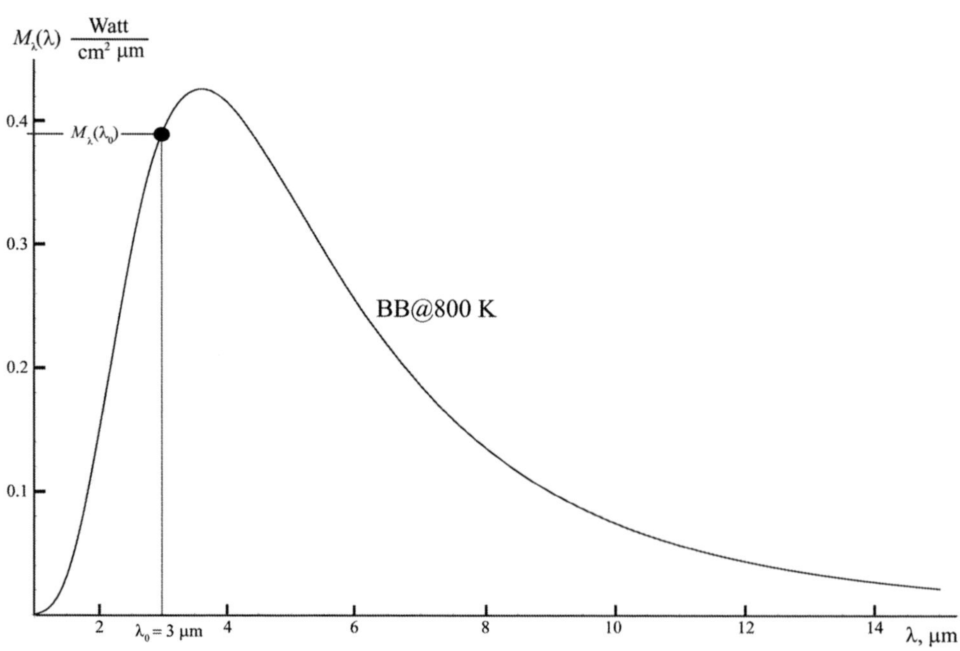

Figure 6.5 A given value of $M_\lambda(\lambda_0)$ is unique to one value of blackbody temperature, shown here for $T = 800$ K.

In a brightness temperature measurement, the received flux scales with $\Delta\lambda$. Thus, the post-sensor SNR is reduced by the narrowband spectral filtering of $\Delta\lambda$ around λ_0, compared to a broadband measurement. For optimum sensitivity, it is beneficial to choose the measurement wavelength in a region where M_λ changes rapidly with temperature (see Section 4.4, Exitance Contrast). Like radiation temperature, a measurement of brightness temperature is biased by incomplete knowledge of emissivity or attenuation. However, these values would only need to be known at the measurement wavelength λ_0 to enable a correction back to the true temperature, using Eq. (6.5).

6.4 Color Temperature

A common definition of color temperature T_c is the temperature of the blackbody that matches the ratio of measured values of M_λ at two wavelengths. Given a ratio of measured spectral exitances at wavelengths λ_1 and λ_2, we can solve for T_c using

$$\frac{M_{\lambda,\mathrm{meas}}(\lambda_1)}{M_{\lambda,\mathrm{meas}}(\lambda_2)} = \frac{\lambda_2^5[e^{hc/\lambda_2 kT_c} - 1]}{\lambda_1^5[e^{hc/\lambda_1 kT_c} - 1]}. \tag{6.6}$$

As seen in Fig. 6.6, a graybody and a blackbody have the same color temperature because the constant ε cancels out in the ratio

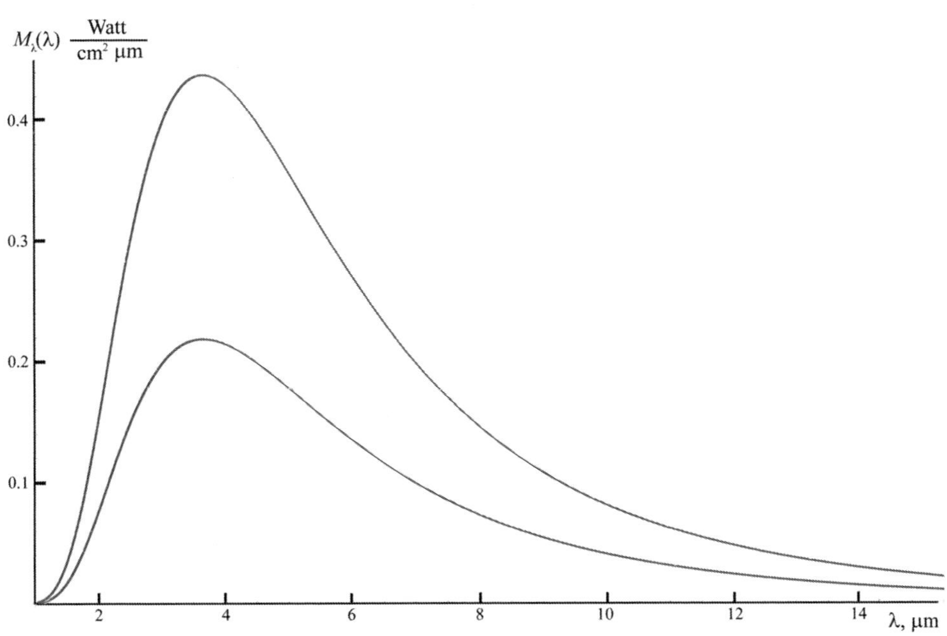

Figure 6.6 The ratio of spectral exitances at any two wavelengths is equal for a blackbody and a graybody, shown here for $T = 800$ K and $\varepsilon = 1.0$ and 0.5.

$$\frac{M_{\lambda,\text{meas}}(\lambda_1)}{M_{\lambda,\text{meas}}(\lambda_2)} = \frac{\varepsilon(\lambda_1)\lambda_2^5[e^{hc/\lambda_2 kT_{\text{true}}} - 1]}{\varepsilon(\lambda_2)\lambda_1^5[e^{hc/\lambda_1 kT_{\text{true}}} - 1]}. \tag{6.7}$$

Thus, color-temperature measurement is robust to an incomplete knowledge of emissivity or attenuation, only if these characteristics are not strongly λ dependent. The same considerations apply as discussed in the context of brightness temperature regarding the spectral locations and spectral bandwidths of the two exitance measurements.

If the source emissivity is strongly wavelength dependent, measurement of M_λ at more than two wavelengths is desirable. One approach uses a spectroradiometer to measure M_λ as a continuous function of λ. As seen in Fig. 6.7, a rotating circular-variable spectral filter is the heart of the instrument, having a narrow spectral bandwidth with a center wavelength that depends on angular position. A curve-fitting algorithm referenced to the Planck function can then be used to estimate the source temperature, noting that the radiation from a real source at a given temperature will be bounded by a blackbody at that temperature under the assumption of thermal equilibrium.

6.5 Practical Temperature Measurements with an IR Camera

Often, we want to know the temperature T of a particular location in an object scene using an IR camera. To do this, we record the signal from various pixels in the detector array. From Eq. (5.10), the pixel signal is a function of the spectral exitance $M_\lambda(\lambda,T)$ of the object at the location of the pixel footprint and the system's spectral responsivity. The spectral exitance at any

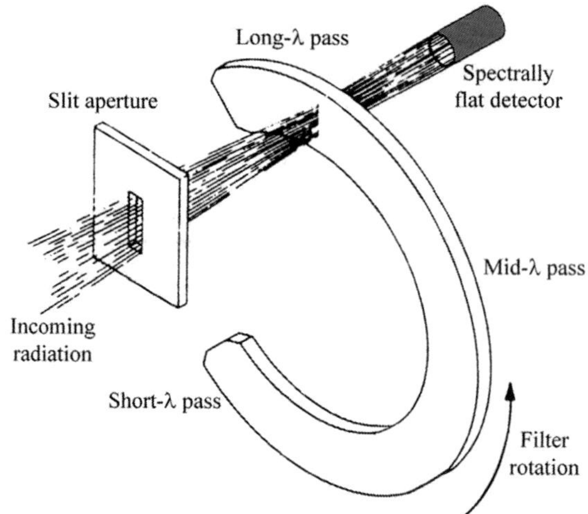

Figure 6.7 A spectroradiometer is based on a rotating circular-variable spectral filter.

location in the object scene is a function of both the local temperature and spectral emissivity.

The signal from a given pixel is the product of the transmittance $\mathcal{T}(\lambda)$ of the optics and any spectral filters, the spectral responsivity of the sensor $\mathcal{R}(\lambda)$, and the throughput of the camera system ($A\Omega/\pi$), yielding

$$\text{Pixel signal} = \frac{A\Omega}{\pi}\int_0^\infty \mathcal{T}(\lambda)\mathcal{R}(\lambda)M_\lambda(\lambda, T)\mathrm{d}\lambda. \tag{6.8}$$

Assuming that the sensor responsivity $\mathcal{R}(\lambda)$ and optics transmittance $\mathcal{T}(\lambda)$ are expressed in energy-based units, we find that

$$\text{Pixel signal} = \frac{A\Omega}{\pi}\int_0^\infty \mathcal{T}(\lambda)\mathcal{R}(\lambda)\varepsilon(\lambda)\frac{2\pi hc^2}{\lambda^5[e^{hc/\lambda kT} - 1]}\mathrm{d}\lambda. \tag{6.9}$$

If the sensor responsivity $\mathcal{R}(\lambda)$ and optics transmittance $\mathcal{T}(\lambda)$ are expressed in photon-based units, we find that

$$\text{Pixel signal} = \frac{A\Omega}{\pi}\int_0^\infty \mathcal{T}(\lambda)\mathcal{R}(\lambda)\varepsilon(\lambda)\frac{2\pi c}{\lambda^4[e^{hc/\lambda kT} - 1]}\mathrm{d}\lambda. \tag{6.10}$$

To estimate T at a particular scene location from the corresponding pixel signal, we make some assumptions. We assume that the throughput of the camera system $A\Omega/\pi$ is the same for all pixels. The transmission of the optics and filters $\mathcal{T}(\lambda)$ is generally given by the manufacturer simply as a passband like "3 to 5 μm," or "8 to 10 μm." We typically have no additional detailed information regarding $\mathcal{T}(\lambda)$ without undertaking relatively complex spectroscopic measurements. Thus, we generally assume $\mathcal{T}(\lambda) \approx 1$ inside the specified passband and zero elsewhere. We use the ideal sensor responsivity (Fig. 5.8) within the passband of the optics. Thus, a photon sensor's spectral response is flat in photon-based units, and a thermal sensor's spectral response is flat in energy-based units. An integration of either watts or photons is taken over the passband from the appropriate version of the Planck equation. If the object's spectral emissivity is known, it can be included in the calculation as in Eqs. (6.9) and (6.10). But generally, the spectral dependence of scene emissivity is not known *a priori*, and any graybody emissivity would combine with the spectrally constant multiplicative throughput factor $A\Omega/\pi$. Either Eq. (6.9) or (6.10) will yield relative signal levels for spatially separated locations within the FOV that correspond to different object temperatures. We will be comparing signal levels in adjacent image regions, so this throughput factor is identical and will cancel out in the computation. The camera system can be calibrated by measuring the signal level for a blackbody source of known temperature. Alternatively, we can use a single-point contact temperature measurement (i.e., a thermocouple) of a high-emissivity location within the FOV. Considering the limited range of sensor linearity, this calibration temperature should be within the anticipated range of scene temperatures.

At this point, we have related a signal level on one pixel or a group of pixels to a known scene temperature. The remainder of the pixels in the scene will scale according to Eq. (6.9) or (6.10), allowing inference of scene temperatures from the measured signal levels. This procedure will allow us to estimate temperatures at various points in the scene. Because of the inherent assumptions and approximations discussed above, the accuracy of these temperature estimates will not be as good as the precision (repeatability), which is determined by the thermal sensitivity of the sensor, expressed as noise-equivalent temperature difference (NETD—discussed in Chapter 9).

Typical sensitivities of 50 mK (1/20°C) for an uncooled bolometer camera are sufficient for many IR-imaging applications. To achieve better thermal sensitivity, you can reduce the frame rate if the camera's frame rate is adjustable. If the frame rate is fixed, you can always acquire several frames of data and average them, which generally increases the SNR by the square root of the number of frames averaged. Either technique assumes a stationary scene over the data-acquisition time.

Bibliography

Inagaki, T., and T. Ishii, "Proposal of quantitative temperature measurements using two-color technique combined with several infrared radiometers having different detection wavelength bands," *Opt. Eng.* **40**, 372–380 (2001) [doi: 10.1117/1.1349218].

Usamentiaga, R., P. Venegas, J. Guerediaga, L. Vega, J. Molleda, and F. G. Bulnes, "Infrared thermography for temperature measurement and non-destructive testing," *Sensors* **14**(7), 12305–12348 (2014).

Chapter 7
Noise in the Detection Process

7.1 Basic Noise Nomenclature

We can express noise in any electrical variable: voltage, current, or power. If we designate a random noise voltage waveform $v_n(t)$, we use three main descriptors for this noise:

$$\text{Mean [V]: } \bar{v}_n = \frac{1}{T}\int_0^T v_n(t)\mathrm{d}t, \tag{7.1}$$

$$\text{Mean-squared [V}^2\text{] } \sigma_v^2 = \frac{1}{T}\int_0^T (v_n(t) - \bar{v}_n)^2\mathrm{d}t, \tag{7.2}$$

and

$$\text{Root-mean-squared(rms) [V] } \sigma_v = \sqrt{\frac{1}{T}\int_0^T (v_n(t) - \bar{v}_n)^2\mathrm{d}t} = v_{\mathrm{rms}}, \tag{7.3}$$

where T is the length of the data record in time units.

As seen in Fig. 7.1, when adding two uncorrelated random waveforms (that is, noise contributions arising from different sources), the mean values of the waveforms add because the means are constant quantities. The peaks and valleys of the fluctuating portion of the waveforms do not line up, and thus the rms voltage of the sum adds *in quadrature*, that is, a linear addition in power rather than in voltage. This operation is expressed as Eq. (7.4):

$$v_{\mathrm{rms,total}} = \sqrt{(v_{\mathrm{rms},1})^2 + (v_{\mathrm{rms},2})^2} \neq v_{\mathrm{rms},1} + v_{\mathrm{rms},2}. \tag{7.4}$$

For example, two independent noise sources, seen in Fig. 7.2, each of 3 μV rms, are added in series. In this case, the total rms noise is $v_{\mathrm{rms,total}} = \sqrt{(3\,\mu\mathrm{V})^2 + (3\,\mu\mathrm{V})^2} = 4.24\,\mu\mathrm{V} \neq 6\,\mu\mathrm{V}$.

This calculation often arises in the context of removing the contribution of amplifier noise (known from specifications or from a separate measurement)

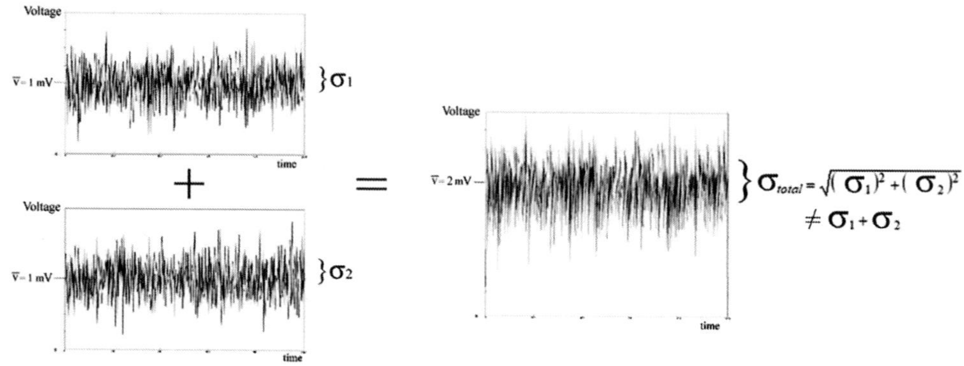

Figure 7.1 Addition of uncorrelated random waveforms in quadrature.

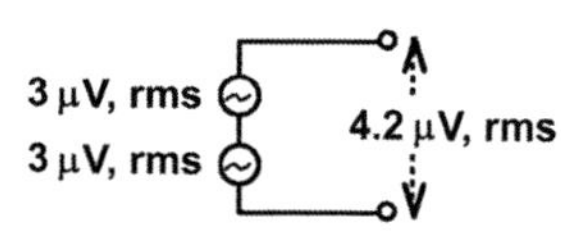

Figure 7.2 Uncorrelated noise sources add in quadrature.

from the measured noise of a detector/amplifier combination, to obtain just the rms detector noise. The total rms noise is $v_{\text{rms,total}} = \sqrt{v_{\text{det}}^2 + v_{\text{amp}}^2}$, yielding the rms detector noise alone as $v_{\text{det}} = \sqrt{v_{\text{rms,total}}^2 - v_{\text{amp}}^2}$.

7.2 Sources of Noise

We must consider many sources of noise in the design of a sensing system, including external interference, amplifier noise, and internally generated noise. External interference caused by other radiation sources can be controlled by optical shields, baffles, or light-tight enclosures to limit the radiation reaching the sensor to only that intended in the measurement. For laboratory characterization of IR systems, the heat signature from human bodies or from lab equipment, reflected from walls (even slightly rough surfaces that are not specular in the visible band can be quite reflective at long wavelengths), can put extraneous radiation into the FOV of a sensor. Often electrical measurements must be made in relatively dense electromagnetic-interference (EMI) environments, with a combination of signals present in various frequency ranges from wi-fi, cellphones, Bluetooth, and laboratory equipment. A typical first step toward mitigating this is the ubiquitous aluminum foil often seen encasing signal-carrying paths in experimental setups. Depending on the frequencies of the strongest interference (perhaps identified using a spectrum analyzer), metallic or MuMetal® foils or wire meshes can be used for shielding. For component and sub-system testing, convenience dictates keeping the region that requires shielding as small as possible, with characterization of larger systems carried out in EMI-screened enclosures if needed. We must remember that all wires act as antennas to some degree. We can reduce interference pickup by keeping the distance between the sensor and the first stage of amplification as short as possible, which can lead to preamplifiers being located inside the dewar of the cooling system to reduce the distance between the sensor and the preamp. Cooling a preamp also reduces its noise contribution.

Amplifier noise can be the limiting factor in a particular experimental setup, especially if a simple amplifier is configured on a protoboard. However, with careful electronic design and packaging, it is generally possible to reduce amplifier noise below the internally generated sensor noise. Commercial amplifiers are specified in terms of their noise floor, most often in units of rms volts/$\sqrt{\text{Hz}}$, which allows for comparison of amplifier performance to the expected signal and noise levels of the sensor. The goal is to have the internally generated sensor noise be the dominant noise term in the quadrature sum of Eq. (7.4).

7.3 Internally Generated Sensor Noise

Internally generated sensor noise typically gets worse with increased sensor temperature. The thermal generation of charge carriers across the bandgap of a semiconductor (the so-called *dark current*) discussed in Section 5.3 has a shot-noise contribution. Thus, a colder detector reduces dark current, which results in less noise. If the sensor noise is white (containing all frequencies in equal amounts), then the total noise power is proportional to bandwidth. In this case, the rms noise voltage is proportional to the square root of the measurement bandwidth Δf.

Empirically, the noise power of a sensor is proportional to the volume of the sensor material. The thickness of the sensor is sufficient for efficient absorption of photons but typically not much more. With the thickness thus fixed, the volume of the sensor is then proportional to the active area (facing the incoming radiation). The measured rms sensor noise is thus generally proportional to the square root of the active area of the sensor, so large-area detectors have more noise than small detectors. In summary, typically the rms noise voltage is proportional to both $\sqrt{A_{\text{det}}}$ and $\sqrt{\Delta f}$. We will watch for these dependencies in our noise expressions. The required bandwidth and sensor area are determined by the mission that the system is designed to perform.

The spatial-resolution requirements of the system will determine the required instantaneous field of view (IFOV), which is related to the detector area and focal length of the optics by IFOV $= \sqrt{A_{\text{det}}}/f$, as seen in Figs. 7.3 and 2.10.

Similarly, from a noise viewpoint, the measurement bandwidth of the sensor should not be larger than the data bandwidth needed to pass the

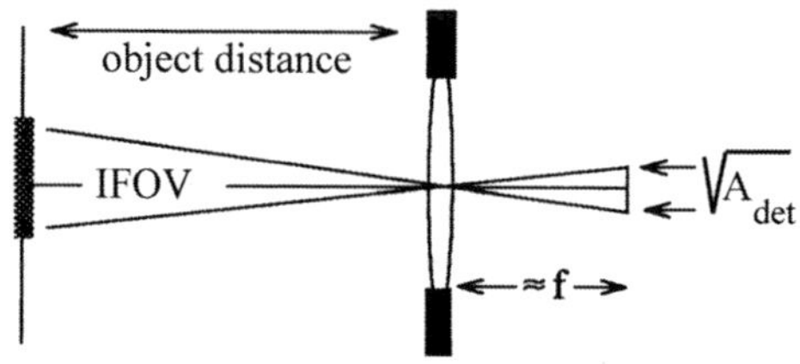

Figure 7.3 Detector area related to IFOV.

desired information. Extra bandwidth (a faster-than-necessary temporal response) will just add more noise. The usual assumption is that sensor noise is white (independent of frequency). This is not strictly true because various noise mechanisms have different frequency contents. However, for most noise calculations, the white-noise assumption is valid within the frequency passband of the system. The bandwidth Δf is related to the sensor response time τ as discussed in Section 5.4, and the rms noise voltage is proportional to $\sqrt{\Delta f}$.

7.4 Noise Power Spectral Density

Electronics noise can be described in terms of its frequency content, which yields insight into the contributions from various noise mechanisms. Power spectral density (PSD) is the mean-squared noise per unit frequency interval. PSD has units of [watt/hertz]. Conceptually, PSD is close to the Fourier transform squared of the noise voltage waveform:

$$\mathrm{PSD}(f) \approx |\mathcal{F}\{v_n(t)\}|^2. \tag{7.5}$$

There is a mathematical subtlety here. It is not possible to directly take a Fourier transform of a random function $v_n(t)$ because the function is not square integrable over the time interval $-\infty < T < \infty$. Thus, the calculation is performed using a finite-length Fourier transform, $V_T(f)$, defined as

$$V_T(f) = \int_{-T/2}^{T/2} v_n(t)\mathrm{e}^{-j2\pi f t}\,\mathrm{d}t. \tag{7.6}$$

The PSD is then defined as

$$\mathrm{PSD}(f) \equiv \lim_{T\to\infty} \left\langle \frac{1}{T}|V_T(f)|^2 \right\rangle, \tag{7.7}$$

where the limit as the data record length $T \to \infty$ indicates that the true PSD of the underlying random process would be realized from the indicated operations on an infinitely long data record. Because we always deal with finite data records in practice, the ensemble average, denoted by $< >$, is necessary for convergence of a stable PSD from an observed set of finite-length random-noise waveforms.[1] The finite-length Fourier transform is performed on each time record by a fast-Fourier-transform (FFT) computation. Because the experimentally acquired time records of the underlying random process are each different in detail, each data record has a slightly different FFT. When the squared FFTs of the individual data records are averaged together, the ensemble average eventually converges toward a stable estimate of the PSD.

The PSD can be physically interpreted as the amount of power that the random-noise waveform $v_n(t)$ delivers to 1-Ω resistor per unit of measurement bandwidth, conceptually shown in Fig. 7.4. The unit-gain narrow-passband filter has a tunable center frequency f_{ref}. As f_{ref} is varied, PSD(f) is the amount of power dissipated in the 1-Ω resistor from the random voltage waveform. The units of PSD(f) can be found from Eqs. (7.5) to (7.7). The units of $V_T(f)$ are [volt × sec] or [volt/hertz]. So, the units of PSD(f) are [1/sec × {volt/hertz}^{2}] = [watt/hertz]. The units of PSD(f)Δf are [watt], which are the same as volt^2 or amp^2 into a 1-Ω load resistor. Then, the units of $\sqrt{\text{PSD}(f)\Delta f}$ are [volt (rms)] or [amp (rms)], and units of $\sqrt{\text{PSD}(f)}$ are [volt/$\sqrt{}$hertz] or [amp/$\sqrt{}$hertz].

The typical PSD plot for a sensor is shown in Fig. 7.5. From that, we can see that different noise sources have different frequency contents. The frequency dependence of the PSD will affect the choice of the upper and lower bandwidth limits for an imager system with the goal of increasing the SNR. The excess noise at low frequencies is $1/f$ noise, which can be partly excluded by *a/c coupling* that rolls off the system response at dc using an electrical filter network. Shot noise and generation–recombination (G/R) noise have a roll-off frequency in the midband range (≈20 kHz – 1 MHz), proportional to the inverse of charge-carrier lifetime. An electrical filter is also applied in the high-frequency range. In a photon sensor, the charge carriers transmit both

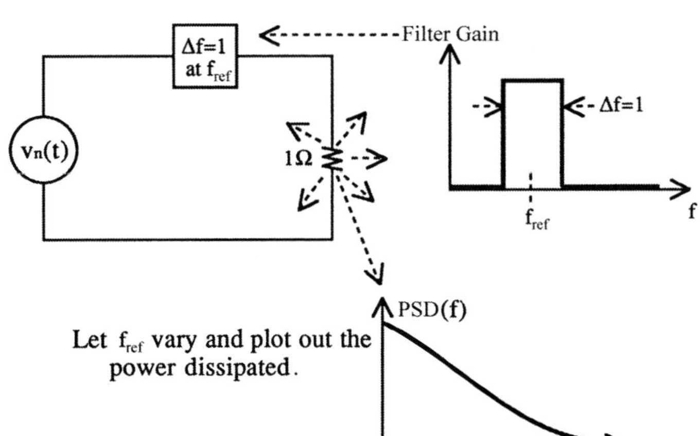

Figure 7.4 Conceptual definition of power spectral density.

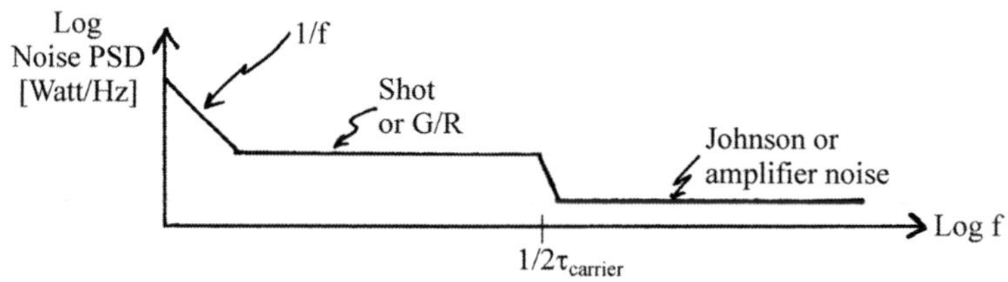

Figure 7.5 Typical log–log plot of noise PSD for a sensor.

signal and noise. Thus, to pass the signal data, the upper cutoff frequency of the electronics should be approximately the inverse of the charge-carrier lifetime $1/2\tau_{carrier}$. Johnson and amplifier noise are usually flat out to frequencies well past this and will bring in extra noise if not excluded by the upper passband limit. A wider bandwidth would include more noise, but not more signal.

The lower- and upper-frequency filters will limit the response of the imager system to those frequencies where the noise PSD is approximately flat. Most system-noise calculations are performed in the midband region between the $1/f$-corner frequency and the carrier roll-off frequency. This justifies the white-noise assumption we typically make in SNR calculations.

Some other items are worth noting. Plots of noise PSD vs frequency are typically made as log–log plots because of the large range of both variables. Displaying the noise PSD on an electronic spectrum analyzer allows for assessment of the various noise contributions, which can be a useful diagnostic laboratory test to verify proper sensor operation. The temporal frequencies contained in the imager signal are associated on a one-to-one basis with spatial frequencies in the image. The proportionality factor between temporal and spatial frequencies has units of a scan velocity [cy/sec] = [cy/mm] × [mm/sec].

7.5 Calculations with White Noise

For white noise, the noise power (area under PSD vs f plot) is proportional to the noise-equivalent bandwidth Δf of the measurement, as seen in Eq. (5.20) and Fig. 5.16. Suppose that a particular noise voltage source is white over the frequency region of interest with PSD = 50 nW/Hz. In $\Delta f = 1$ Hz, this voltage source has 50 nW of noise power. In $\Delta f = 10$ Hz, it has 500 nW of noise power, and in $\Delta f = 500$ kHz, the source has 25 mW of noise power. We can specify the noise-voltage spectrum or noise-current spectrum (for 1-Ω load) as the square root of the PSD, with units of [V/$\sqrt{\text{Hz}}$] or [A/$\sqrt{\text{Hz}}$]. So, for white noise, the rms noise voltage σ_v or current σ_i is proportional to $\sqrt{\Delta f}$. For example, suppose that a noise source is measured to have 7.5 μV rms noise in a bandwidth $\Delta f = 50$ Hz. It has a noise-voltage spectrum of $7.5 \times 10^{-6}/\sqrt{50} = 1.1 \times 10^{-6}$ V$\sqrt{\text{Hz}}$. Over a bandwidth $\Delta f = 500$ kHz, this noise source will yield an rms noise voltage of 1.1×10^{-6} V/$\sqrt{\text{Hz}} \times \sqrt{500,000\ \text{Hz}} = 7.5 \times 10^{-4}$ V.

Consider passing the sensor current (which has additive white noise) through a low-pass filter with a given noise-equivalent bandwidth. If we assume that the filter neither absorbs nor accumulates charge carriers, the filter will not change the mean level of the sensor current. However, because of its finite response time, the filter will attenuate the high frequencies contained in the sensor current. If the input noise is white (contains all frequencies) then the noise-equivalent bandwidth of the filter determines how much noise power passes through the filter. A wider bandwidth (shorter response time) passes more noise power, and a narrower bandwidth (longer response time) passes less noise power. For any white-noise current, the standard deviation of the current fluctuations $\sigma_i \propto \sqrt{\Delta f} = 1/\sqrt{2\tau}$, so using a measurement circuit with a longer response time yields less current noise. We visualize this dependence in Fig. 7.6, where the mean current level is held constant while the response time of the low-pass filter is varied. Increasing the response time by a factor of 4 reduces σ_i by a factor of 2.

7.6 Shot Noise

Shot noise arises because charge carriers are quantized; we cannot divide an electron in half. Shot noise occurs in photon detectors in which the sensor current flows across a potential-energy barrier. One example of this situation

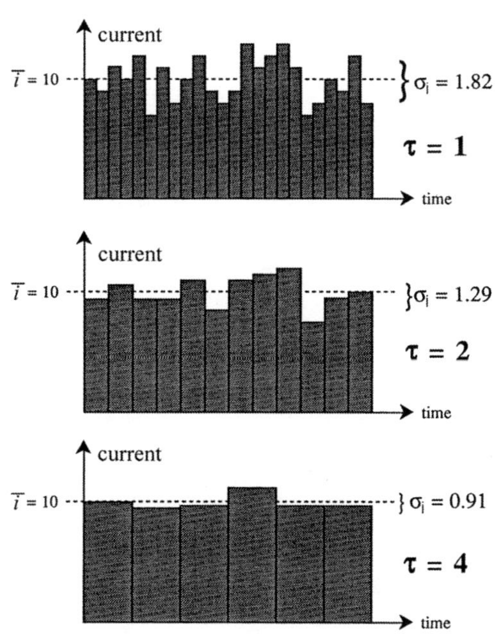

Figure 7.6 White noise as a function of measurement time for constant mean current.

is a photovoltaic sensor operating in a mode where current is flowing (short-circuit, reverse-bias, or forward-bias conditions). The p–n junction is the potential-energy barrier, which causes immediate separation of hole–electron pairs once the photons are detected. Other examples are external photo-effect devices such as a photomultiplier tube, where the electrons see a potential-energy barrier between the cathode and vacuum emission and Schottky-barrier detectors. In these cases, the generation of carriers is random, according to photon-arrival times. But once a carrier is generated, the carrier lifetime is deterministic, determined by transit-time considerations rather than by a statistically determined carrier lifetime.

At frequencies below the inverse of the charge-carrier lifetime, the rms noise current σ_i [Amp] is white noise with a value of

$$\sigma_i = \sqrt{2q\bar{i}\Delta f}, \tag{7.8}$$

where q is the electron charge (1.6×10^{-19} coul), $\bar{i}$ is mean value of current, and $\Delta f \approx 1/2\tau$ is measurement bandwidth. As noted previously, a faster measurement-circuit response leads to a larger bandwidth and consequently more noise. The units are seen to be consistent: $[\text{coul} \times \text{coul}/\text{sec} \times 1/\text{sec}]^{1/2} = [\text{amp}]$.

7.6.1 Development of the shot-noise expression

We will develop Eq. (7.8) using Fig. 7.7 under the assumption that electrons are particles that obey Poisson statistics, which implies that the mean equals the variance. The current during any measurement interval is $i = nq/\tau$, where n is the number of electrons in that interval. Similarly, the average current is $\bar{i} = \bar{n}q/\tau$. The variance of the current fluctuations is then $\sigma_i^2 = q^2\langle(n-\bar{n})^2\rangle/\tau^2 = q^2\sigma_n^2/\tau^2$. Now we invoke the assumption of Poisson statistics, which means that the variance in the number of electrons equals the mean number, $\sigma_n^2 = \bar{n}$, which yields (using $\Delta f = 1/2\tau$)

$$\sigma_i^2 = \frac{q^2}{\tau^2}\bar{n} = \frac{q^2}{\tau^2}\frac{\bar{i}\tau}{q} = \frac{q\bar{i}}{\tau} = 2\Delta f\, q\bar{i}, \tag{7.9}$$

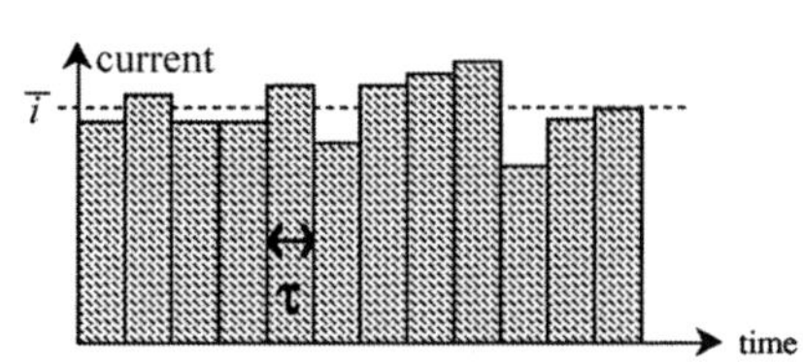

Figure 7.7 Development of the shot-noise expression.

which yields Eq. (7.8). Shot-noise power is independent of frequency out to $f \approx 1/2\tau_{carrier}$. Beyond that, as seen in Fig. 7.5, the shot-noise power decreases with frequency according to Eq. (7.10):

$$\sigma_i^2 = \frac{2q\bar{i}\Delta f}{1 + (2\pi\tau f)^2}. \tag{7.10}$$

7.6.2 What process generates the current?

Per Eq. (7.8), the rms shot noise is proportional to the square root of the mean current, as illustrated in Fig. 7.8. But, from a signal-to-noise ratio perspective, the question is: what process is primarily responsible for generating the current?

If the sensor current is generated primarily by signal photons, then a bigger current is better from an SNR point of view. Because the rms noise increases only as the square root of mean current, the SNR improves with increased current. As seen in the left diagram of Fig. 7.9, if we assume that the mean current is the signal, the SNR is better when the mean current is 20 (SNR = 20/2.57 = 7.8) than when the mean current is 5 (SNR = 5/1.29 = 3.9), even though the fluctuations are larger for a bigger current.

However, as seen in the right diagram of Fig. 7.9, if the signal current is small compared to the current caused by background photo-generated

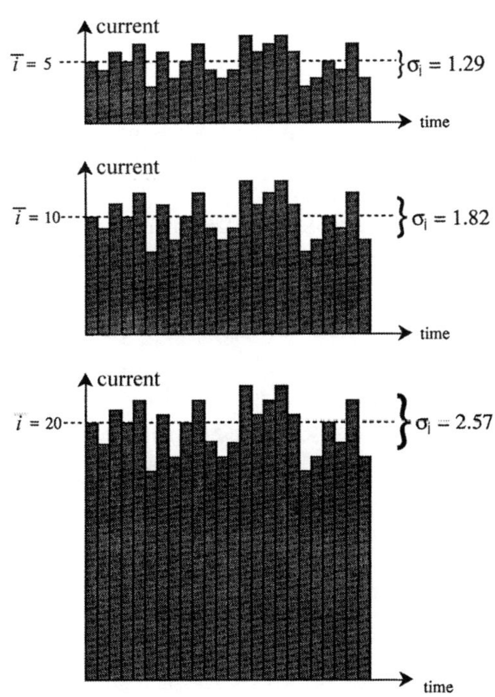

Figure 7.8 Shot noise vs mean current level for a constant measurement bandwidth.

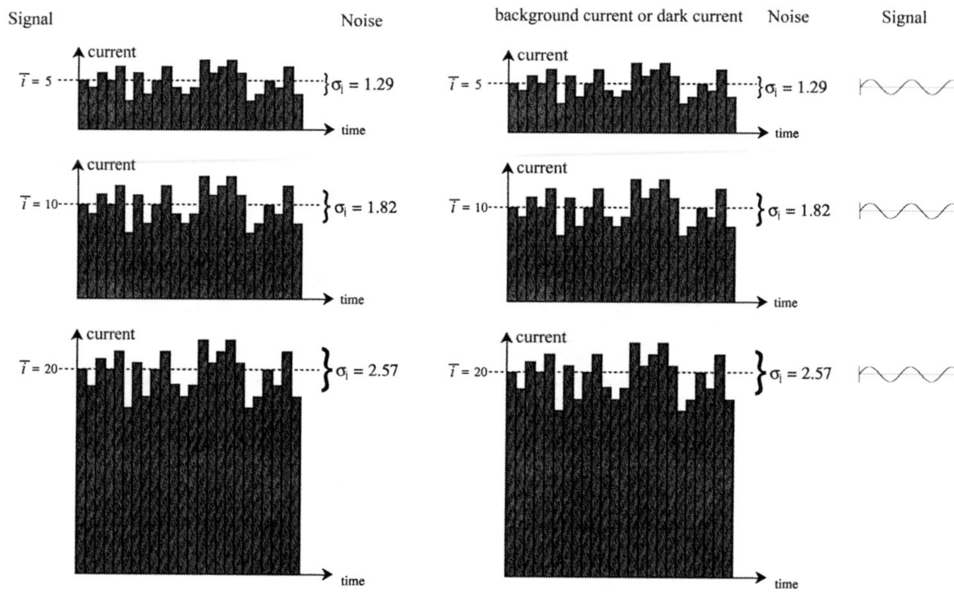

Figure 7.9 (left) Signal-shot-noise limit. (right) Background or dark-current shot-noise limit.

electrons or by dark-current electrons, then a bigger current is worse from an SNR point of view. The rms noise increases as the square root of mean current, and a small signal current is difficult to detect in the presence of the shot-noise fluctuation on the non-signal currents. In this case, the SNR decreases for increased current.

7.7 SNR in the Signal-Shot-Noise Limit

We will consider two scenarios to illustrate the signal-shot-noise limit. The result for both is that the SNR is equal to the square root of the number of detected photons in a measurement interval. In the first scenario, we consider the signal and noise to be expressed in terms of the current flow [amp]. According to Eq. (5.2), in a photon sensor, the current generated from signal photons is $i_{\text{sig}} = \phi_{q,\text{sig}}\eta q$, where η is the quantum efficiency [electrons/photon] and $\phi_{q,\text{sig}}$ is the signal photon flux. Similarly $i_{\text{bkg}} = \phi_{q,\text{bkg}}\eta q$ is the background current, where $\phi_{q,\text{bkg}}$ is the background photon flux. Dark current i_{dark} is generated by either thermally excited carriers or a reverse-bias current. In general, we express the SNR as

$$\text{SNR} = \frac{i_{\text{sig}}}{\sigma_i} = \frac{\phi_{q,\text{sig}}\eta q}{\sqrt{2q\bar{i}\Delta f}}. \tag{7.11}$$

As seen in Eq. (7.12), the mean current $\bar{i}$ is the sum of signal current, background current, and dark current. In the signal-shot-noise limit of

Eq. (7.13), we assume that $i_{\text{sig}} \gg i_{\text{bkg}}$ (which is difficult to achieve in an IR-imaging context) and that $i_{\text{sig}} \gg i_{\text{dark}}$. Under those assumptions, the dominant noise contribution in the quadrature sum is the shot noise on the signal. This is the ideal situation, not often achieved in practice except in coherent heterodyne receivers (e.g., in laser radars) and with photomultiplier tubes used in the visible or near-IR bands:

$$\text{SNR} = \frac{\phi_{q,\,\text{sig}}\eta q}{\sqrt{2q\left(\phi_{q,\text{sig}}\eta q + \phi_{q,\text{bkg}}\eta q + i_{\text{dark}}\right)\Delta f}} \tag{7.12}$$

and

$$\text{SNR} = \frac{\phi_{q,\text{sig}}\eta q}{\sqrt{2q\left(\phi_{q,\text{sig}}\eta q\right)\Delta f}} = \sqrt{\phi_{q,\text{sig}}\eta\tau}. \tag{7.13}$$

From Eq. (7.13), we see that the signal-shot-noise-limited SNR is equal to the square root of the number of detected photons in a measurement interval.

In the second scenario, we consider the signal and noise to be expressed in terms of the number of electrons accumulated in a measurement interval. This is the case for charge accumulation in the potential wells of a charge-coupled-device FPA. In the context of photon detection, a photon either liberates an electron or it does not. Quantum efficiency η is an average yield of electrons/photon, defined statistically. Photons, like electrons, are particles that obey Poisson statistics, where the variance equals the mean. When we accumulate electrons over a measurement interval τ, the mean number of electrons accumulated over a time τ is

$$\bar{n} = \eta\phi_{q,\text{sig}}\tau. \tag{7.14}$$

From Poisson statistics, assuming that other noise sources are negligible, consistent with the signal-shot-noise limit, we find that

$$\sigma_n^2 = \bar{n} = \eta\phi_{q,\text{sig}}\tau. \tag{7.15}$$

Hence, we can write SNR as

$$\text{SNR} = \frac{\bar{n}}{\sigma_n} = \frac{\eta\phi_{q,\text{sig}}\tau}{\sqrt{\eta\phi_{q,\text{sig}}\tau}} = \sqrt{\eta\phi_{q,\text{sig}}\tau}. \tag{7.16}$$

For the carrier-accumulation scenario, the measurement time changes both the signal and noise levels, but the overall SNR result is identical to that obtained previously for shot-noise currents: the SNR in the signal-shot-noise limit is the square root of the number of detected photons in a measurement interval.

The statistical nature of photon arrivals can be most clearly seen when the photon flux levels are low (and the other noises are negligible so that you are still in the signal-shot-noise limit). Consider the series of signal-shot-noise-limited images in Fig. 7.10, where the photon-flux level is constant while the integration time is varied. SNR is seen to increase as the number of detected photons increases.

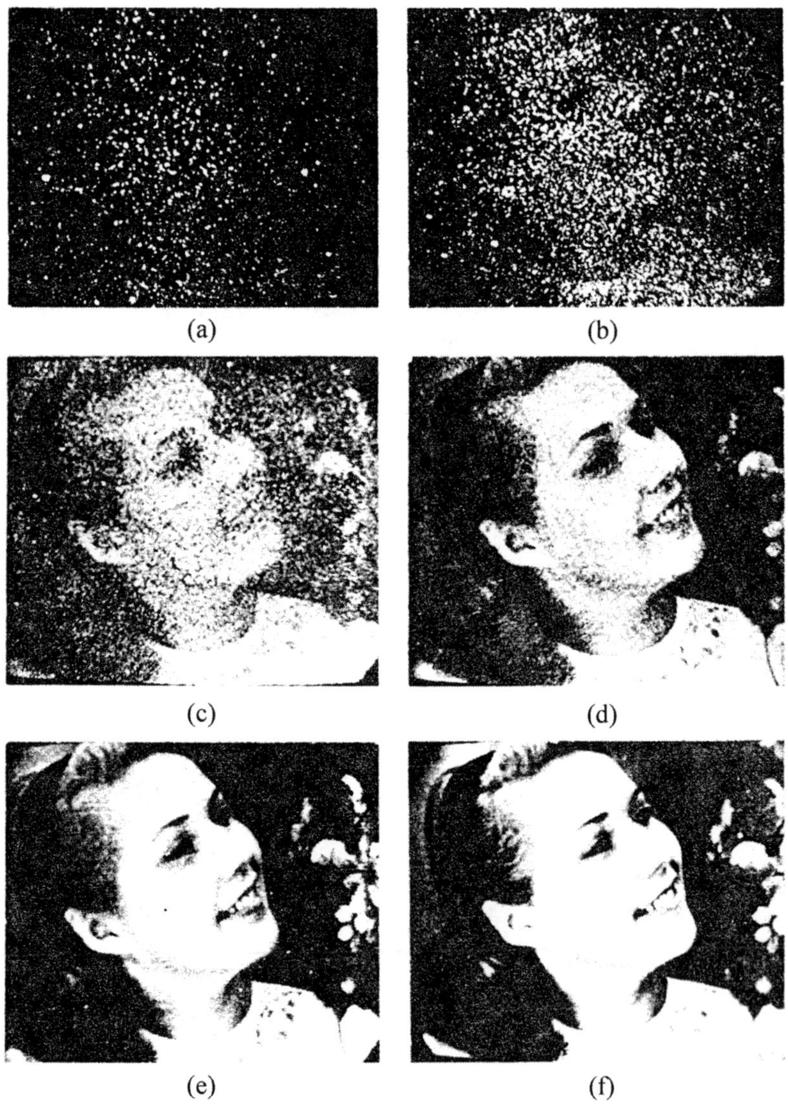

Figure 7.10 Series of signal-shot-noise-limited images where the photon flux level is held constant and the integration time is varied: (a) contains 3×10^3 detected photons, (b) contains 1.4×10^4 detected photons, (c) contains 9.3×10^4 detected photons, (d) contains 7.6×10^5 detected photons, (e) contains 3.6×10^6 detected photons, and (f) contains 2.8×10^7 detected photons. (Adapted from Ref. 2.)

7.8 SNR in the Background-Shot-Noise Limit

When a weak signal source is detected against a strong background (such as an extended background of 300 K, as seen in Fig. 7.11), current is generated by both signal and background photons. We assume that $i_{\text{bkg}} \gg i_{\text{sig}}$ and $i_{\text{bkg}} \gg i_{\text{dark}}$ so that the dominant noise contribution in the quadrature sum of Eq. (7.12) is the shot noise on the background. This is a common situation in the context of IR imaging. The background shot-noise fluctuation remains, even if the background current is subtracted from the mean current $\bar{i}$. Reducing the number of background photons reduces this fluctuation and increases SNR until the noise floor is determined by some other noise source, which will typically be Johnson noise for low-background operation.

Beginning with Eq. (7.12) for the general-case SNR,

$$\text{SNR} = \frac{\phi_{q,\text{sig}}\eta q}{\sqrt{2q\left(\phi_{q,\text{sig}}\eta q + \phi_{q,\text{bkg}}\eta q + i_{\text{dark}}\right)\Delta f}}, \tag{7.17}$$

we apply the conditions $i_{\text{bkg}} \gg i_{\text{sig}}$ and $i_{\text{bkg}} \gg i_{\text{dark}}$ for the background shot-noise limit, yielding

$$\text{SNR} = \frac{\phi_{q,\text{sig}}\eta q}{\sqrt{2q\left(\phi_{q,\text{bkg}}\eta q\right)\Delta f}} = \frac{\phi_{q,\text{sig}}\sqrt{\eta\tau}}{\sqrt{\phi_{q,\text{bkg}}}}. \tag{7.18}$$

Equation (7.18) shows that when the limiting noise is the shot noise on the background, reducing the background photon flux $\phi_{q,\text{bkg}}$ increases the SNR. The background photon flux depends on the temperature of the background and on the solid angle of the background that is viewed by the sensor. In laboratory measurement scenarios, a reduction of background flux can sometimes be accomplished by judicious placement of sponges saturated with liquid nitrogen within the FOV of the sensor. For imagers deployed in the field, we obviously do not have direct control of the background temperature. For sensor systems on a satellite platform, the background level will vary depending on whether the FOV of the sensor includes the warm earth or the cold sky.

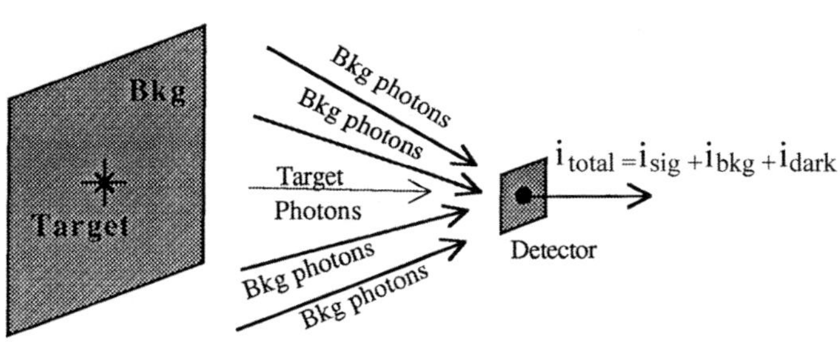

Figure 7.11 Scenario for the background-shot-noise limit.

As discussed in Section 5.3, IR photodetectors are cryogenically cooled to reduce the thermally generated dark current. To accomplish this, the sensor is often housed in a dewar that keeps the detector at a low temperature. Spectral filtering can be used to enhance target detectability by exploiting differences in the spectral-flux distribution between the target and background. This filtering reduces the number of background photons by limiting the wavelength range of the received flux from the background. As seen in Fig. 7.12, a spectral filter can be placed inside the dewar in front of the sensor, allowing the filter to be cooled to the detector temperature. Cooling the spectral filter is beneficial because any filter has some residual absorption and, by Kirchhoff's law, the filter will radiate with an emissivity equal to its absorption. The proximity of the filter to the detector makes this important, even for low levels of absorption because the filter will subtend a large solid angle when viewed from the sensor.

Inside the dewar is a surface called a cold shield that determines the angle through which the sensor views the scene, as seen in Fig. 7.12. Ideally, the aperture stop of the system should be imaged onto the plane of the cold shield, and the cold shield should be the same size as the image of the aperture stop at that location. This arrangement allows only scene radiation transmitted by the optical system to reach the detector. In that case, the cold shield is called a *cold stop*. If the diameter of the cold shield is larger than the image of the aperture stop at that plane, some warm radiation from uncooled opto-mechanical housings will reach the detector, which will contribute to background shot noise.

According to Eq. (7.8), the rms shot noise is proportional to the square root of the mean current flowing from any source. For extended-source imaging (either target or background), the received flux is proportional to the area of the detector for a given sensor-plane irradiance. Thus, the usual area and bandwidth dependences discussed in Section 7.3 apply, with the rms shot noise proportional to $\sqrt{A_{det}}$ (for extended sources) and $\sqrt{\Delta f}$ (for frequencies in the flat region of the PSD, as seen in Fig. 7.5). For point-source imaging,

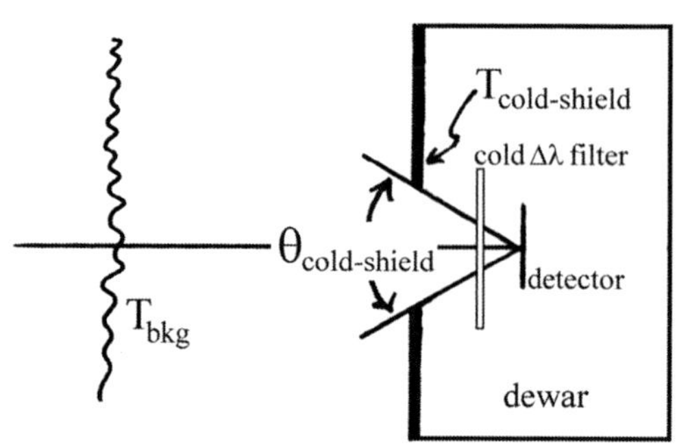

Figure 7.12 Reduction of background photon flux.

the detector should not be larger than the sensor-plane spot size. Extra sensor area will contribute more noise, proportional to $\sqrt{A_{\text{det}}}$, while not collecting additional flux from a point-source target.

7.9 Generation–Recombination Noise

Shot noise occurs in diodes and other devices with a potential-energy barrier, for which generation is a random process, while the recombination process is deterministic, controlled by carrier-transit times. In devices such as photoconductors without a potential-energy barrier, both generation and recombination are random processes. In a photoconductor, once a carrier is generated, it travels under the influence of an applied electric field. The charge carrier lasts until recombination occurs at a random time that is consistent with the mean carrier lifetime τ_{carrier}. For a given photon-flux level, the rms generation–recombination (G-R) noise is $\sqrt{2}$ larger than for shot noise, as there is a shot-noise contribution from both the generation and recombination processes, which add together in quadrature, resulting in

$$\sigma_{i,\text{G-R}} = \sqrt{2q\bar{i}\Delta f + 2q\bar{i}\Delta f} = 2\sqrt{q\bar{i}\Delta f} = \sqrt{2}\,\sigma_{i,\text{shot}}. \tag{7.19}$$

We will see in Chapter 8 that photovoltaic detectors have a $\sqrt{2}$ better limiting performance than do photoconductors, and G-R noise is the origin of that difference.

In Section 5.6.1.2, we introduced the photoconductive gain G, which multiplies the signal current responsivity in Eq. (5.49) as well as the $\bar{i}$ in Eq. (7.19). By controlling the bias voltage, G can be adjusted so that the G-R noise is larger than other sources of noise in the quadrature sum of noise terms. This leads to an SNR that is proportional to $\sqrt{G}$ until the SNR ultimately decreases because of Joule heating.

7.10 Johnson Noise

At any temperature above absolute zero, the charge carriers in the material of a resistor have thermal energy that is manifested as motion of the carriers. Because of this thermal motion, a (zero-mean) rms voltage or current is seen across the terminals of the resistor. A real resistor at $T > 0$ can be modeled as an ideal noise-free resistor of the same resistance value, combined either in series with an rms noise voltage or in parallel with an rms noise current. The Thevenin model of Fig. 7.13 is the series combination of a resistance R in series with rms noise voltage,

$$\sigma_v = \sqrt{4kTR\Delta f}, \tag{7.20}$$

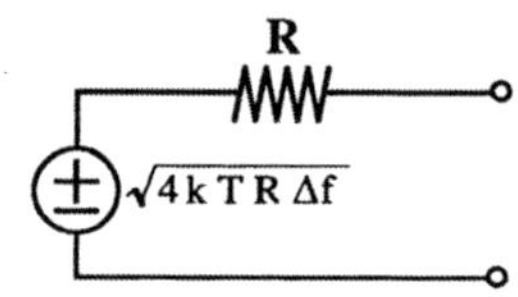

Figure 7.13 Thevenin model of Johnson noise.

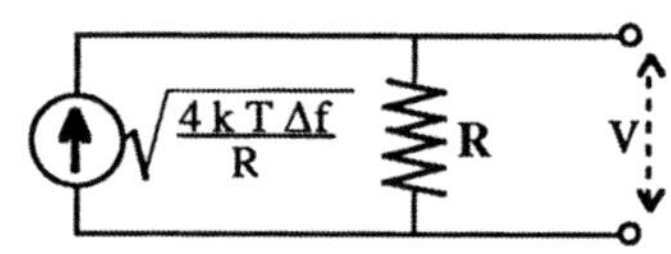

Figure 7.14 Norton model of Johnson noise.

while the Norton model of Fig. 7.14 is the parallel combination of a resistance R in parallel with an rms noise current:

$$\sigma_i = \sqrt{\frac{4kT\Delta f}{R}}. \tag{7.21}$$

These models are two-terminal equivalents, which means that if you connect a load resistance R_L across the terminals of either circuit, the same noise power P_L is dissipated in the load resistor, as seen in Fig. 7.15.

7.10.1 Johnson-noise units

Consider Johnson noise in terms of the noise-voltage spectrum $\sqrt{4kTR}$ with units of [V/$\sqrt{\text{Hz}}$]. We can verify that the units are consistent: [J/K K V/A]$^{1/2}$ = [J V/A V/V]$^{1/2}$ = [J V^2/W]$^{1/2}$ = V/Hz$^{1/2}$. Similarly, the noise-current spectrum $\sqrt{4kT/R}$ has units of [A/$\sqrt{\text{Hz}}$], which are also consistent: [J/K K A/V]$^{1/2}$ = [J A/V A/A]$^{1/2}$ = [J A^2/W]$^{1/2}$ = A/Hz$^{1/2}$. If we multiply these two expressions, we find that the Johnson-noise PSD depends only on temperature, not on resistance. PSD = $4kT$ [J/K K] = [J] = [J/s s] = [W/Hz]. A Johnson-noise rule of thumb is that a 50-Ω resistor at room temp (300 K) has a voltage spectrum of 0.9 nV/Hz$^{1/2}$. In a bandwidth Δf = 500 kHz, the rms Johnson noise voltage in this case is $\sigma_v = 0.64$ μV.

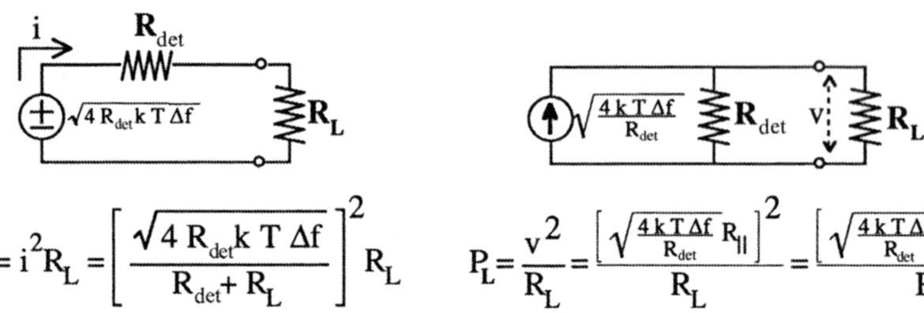

$$P_L = i^2 R_L = \left[\frac{\sqrt{4R_{det}kT\Delta f}}{R_{det}+R_L}\right]^2 R_L \qquad P_L = \frac{v^2}{R_L} = \frac{\left[\sqrt{\frac{4kT\Delta f}{R_{det}}}\,R_{||}\right]^2}{R_L} = \frac{\left[\sqrt{\frac{4kT\Delta f}{R_{det}}}\,\frac{R_L R_{det}}{R_L+R_{det}}\right]^2}{R_L}$$

Figure 7.15 Two-terminal equivalent models for Johnson noise.

7.10.2 Johnson-noise calculation example: two parallel resistors at different temperatures

Often the detector is cooled to 77 K, while the load resistance, corresponding to the input impedance of the preamplifier, is at 300 K. The detector resistance R_d and the load resistance R_L each contribute to Johnson noise. When combined in parallel, the noise-free resistors add as usual, but the zero-mean rms noises add in quadrature, as shown in Fig. 7.16, with the result of

$$i_{\text{total}} = \sqrt{4k\Delta f\left[\frac{T_1}{R_1} + \frac{T_2}{R_2}\right]}. \tag{7.22}$$

7.10.3 SNR in the Johnson-noise limit

For a photovoltaic sensor, we can add a Johnson-noise term to Eq. (7.12) for SNR:

$$\text{SNR} = \frac{\phi_{q,\text{sig}}\eta q}{\sqrt{\left[2q\left(\phi_{q,\text{sig}}\eta q + \phi_{q,\text{bkg}}\eta q + i_{\text{dark}}\right) + \frac{4kT}{R_{\text{det}}}\right]\Delta f}}. \tag{7.23}$$

Particularly for situations with low backgrounds (e.g., space applications that do not have 300 K earth in the FOV), the Johnson-noise term can dominate the quadrature sum, yielding the SNR expression

$$\text{SNR} = \frac{\phi_{q,\text{sig}}\eta q}{\sqrt{\frac{4kT}{R_{\text{det}}}\Delta f}}. \tag{7.24}$$

The Johnson noise is independent of photogeneration processes, and thus the SNR in the Johnson-noise limit is directly proportional to quantum efficiency η, rather than $\sqrt{\eta}$, as was the case for the signal-shot-noise-limited and background-shot-noise-limited SNRs of Eqs. (7.13) and (7.18), respectively.

7.10.4 Johnson noise root-area and root-bandwidth dependence

Since the Johnson noise PSD is white over the frequency range of interest, the rms noise is proportional to $\sqrt{\Delta f}$. To find the dependence of rms Johnson

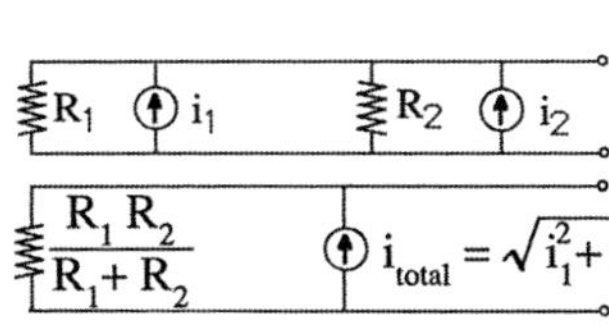

Figure 7.16 Combination of two resistors at different temperatures.

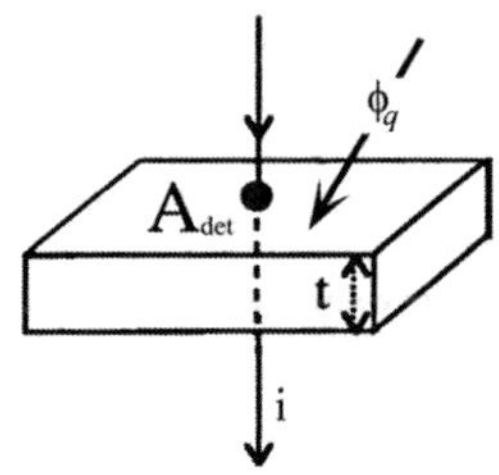

Figure 7.17 Sensor-area dependence of Johnson noise.

noise on the sensor area, consider the configuration of Fig. 7.17, where the sensor is illuminated through the top (transparent) electrode. Given ρ, the resistivity of the sensor material, the detector resistance, is

$$R_d = \rho t / A_{det}. \tag{7.25}$$

For the configuration shown in Fig. 7.17, the rms Johnson-noise current is proportional to $\sqrt{A_{det}}$:

$$\sigma_i = \sqrt{\frac{4kT\Delta f}{R_d}} = \sqrt{\frac{4kT\,\Delta f\,A_{det}}{\rho t}}. \tag{7.26}$$

7.11 1/f Noise

Another noise contribution is $1/f$ noise, which increases at low frequencies, as seen in the PSD plot of Fig. 7.5. This noise is often seen in measurements as a slow drift over time in the observed response. The main causes[3] of $1/f$ noise in IR sensors are surface traps, material defects, poor surface passivation, and poor metal–semiconductor contacts. The $1/f$ noise is typically the dominant noise out to frequencies of a few hundred hertz. If a significant amount of $1/f$ noise is present, its effects can be seen out to several kilohertz. The functional form of the $1/f$ noise PSD is

$$\text{PSD}(f)\ [\text{W/Hz}] \propto (\bar{i})^2/f, \tag{7.27}$$

where $\bar{i}$ is the average current flowing in the sensor. The $1/f$ noise is always present when a sensor current flows, as in bolometers and photoconductors. The $1/f$ noise can be eliminated by not allowing a current flow though the sensor. This situation can be achieved in the open-circuit mode of a photovoltaic, allowing a voltage to form across the device but without current flow. For an IR imager, the transfer function of the electronics typically does not respond all the way to dc (*ac coupled*), to avoid the high $1/f$-noise region at low frequencies. When we measure sensor performance in a laboratory environment, the radiation source is typically modulated by a mechanical chopper wheel at a frequency of a few kilohertz to avoid the effects of $1/f$ noise.

7.12 Thermal-Fluctuation Noise

Thermal-fluctuation noise is the ultimate limit for thermal-sensor performance. The temperature of the sensor will fluctuate because of the radiative exchange between the sensor and the environment. In general, this will yield a poorer limiting SNR for thermal sensors than for photon sensors. The radiative exchange occurs regardless of the sensing mechanism employed but, because photon sensors are not measuring temperature, the performance of a photon sensor is not limited by thermal-fluctuation noise. A body at temperature T will radiate power according to the Planck equation. However, given that the radiation mechanism involves the emission of discrete quanta (photons), there will be some unavoidable fluctuation in the radiated power. The mean-square noise power of this radiation in a bandwidth Δf is found to be[4]

$$\overline{(\Delta\phi)^2} = \varepsilon 8 A_{\mathrm{det}} \sigma k T^5 \Delta f. \tag{7.28}$$

The units of Eq. (7.28) are consistent. Substituting the units of the various terms, we find that $[\overline{(\Delta\phi)^2}] = \mathrm{cm}^2 \frac{\mathrm{watt}}{\mathrm{cm}^2\mathrm{K}^4} \frac{\mathrm{joule}}{\mathrm{K}} \mathrm{K}^5 \frac{1}{\mathrm{sec}} = \mathrm{watt}^2$. Equation (7.28) applies to both the absorbed and emitted radiation streams in the radiative-exchange process between the sensor and the background, so we add the mean-square noise powers of each process directly:

$$\overline{(\Delta\phi)^2} = \alpha 8 A_{\mathrm{det}} \sigma k T_{\mathrm{bkg}}^5 \Delta f + \varepsilon 8 A_{\mathrm{det}} \sigma k T_{\mathrm{det}}^5 \Delta f. \tag{7.29}$$

By Kirchhoff's law, we can set $\alpha = \varepsilon$ for the detector, yielding

$$\overline{(\Delta\phi)^2} = \varepsilon 8 A_{\mathrm{det}} \sigma k (T_{\mathrm{det}}^5 + T_{\mathrm{bkg}}^5) \Delta f. \tag{7.30}$$

As a point of reference, when T_{det} and T_{bkg} are both at 300 K, a unit-emissivity sensor area of 1 mm^2 will have an rms noise power of $\sqrt{\overline{(\Delta\phi)^2}} = 5.5 \times 10^{-12}$ W. To express the SNR in the thermal-fluctuation limit, we take the signal as equal to the absorbed power and the noise as equal to the rms noise power:

$$\mathrm{SNR} = \frac{\phi_{\mathrm{incident}} \varepsilon_{\mathrm{detector}}}{\sqrt{\overline{(\Delta\phi)^2}}} = \frac{\phi_{\mathrm{incident}} \varepsilon_{\mathrm{det}}}{\sqrt{\varepsilon_{\mathrm{det}} 8 A_{\mathrm{det}} \sigma k (T_{\mathrm{det}}^5 + T_{\mathrm{bkg}}^5) \Delta f}} = \frac{\phi_{\mathrm{incident}} \sqrt{\varepsilon_{\mathrm{det}}}}{\sqrt{8 A_{\mathrm{det}} \sigma k (T_{\mathrm{det}}^5 + T_{\mathrm{bkg}}^5) \Delta f}}. \tag{7.31}$$

Equation (7.31) is the fundamental performance limit for thermal sensors. We can see that a better SNR performance is obtained when either the detector or the background is cold. However, because of the fifth-power dependence, cooling the sensor alone provides only a $\sqrt{2}$ increase in SNR. Thus, only for cold backgrounds is there a more significant benefit from cooling the sensor. Thermal sensors can be cooled to reduce the thermal-fluctuation noise, but only if the sensor has a cold background in its FOV. This is a different reason than the cooling of photon sensors referred to in Section 5.3, where the cooling is used to avoid thermal generation of charge carriers across the bandgap in a semiconductor.

References

1. J. Bendat and A. Piersol, *Random Data: Analysis and Measurement, 4th Edition*, Wiley, New York (2011).
2. A. Rose, "Quantum and noise limitations of the visual process," *JOSA* **43**(9), 715–716 (1953).
3. M. A. Kinch, *Fundamentals of Infrared Detector Materials*, SPIE Press, Bellingham, Washington (2007) [doi: 10.1117/3.741688].
4. P. W. Kruse, L. D. McGlauchlin, and R. B. McQuistan, *Elements of Infrared Technology*, Wiley, New York (1962).

Bibliography

Ambrózy, A., *Electronic Noise*, McGraw Hill, New York (1983).

Davenport, W. B., Jr., and W. L. Root, *An Introduction to Random Signals and Noise*, McGraw Hill, New York (1958).

Kogan, S., *Electronic Noise and Fluctuations in Solids*, Cambridge University Press (1996).

Chapter 8
Detector Sensitivity Figures of Merit

8.1 Noise-Equivalent Power (NEP)

We want to specify the noise performance of a detector in a way that allows for prediction of the SNR that we will achieve when a given amount of power falls on the detector. In optics, the SNR is defined as the peak signal voltage divided by the rms noise voltage. This is different from the usual convention in radio-frequency systems, where SNR is defined as signal power divided by noise power. As seen in Fig. 8.1, the noise-equivalent power (NEP) is the amount of power [W] incident on the detector that would produce an output voltage equal to the rms noise voltage (assuming a linear responsivity as seen in the figure). So, NEP is the input flux required to produce SNR = 1. A smaller NEP means better sensitivity (higher SNR per watt of power on the detector). We would prefer a sensor with a nanowatt NEP than one with a milliwatt NEP.

From Fig. 8.1, the responsivity curve is the output (volts or amperes) per watt of input power. As the flux on the detector decreases, the sensor's output decreases until the noise floor of the sensor is reached, at which point the output stops decreasing as input flux decreases. The NEP is the flux on the detector required to make the output of the detector equal to the rms value of the noise. Using these relations, there are various ways to express NEP, seen in Eqs. (8.1) and (8.2):

$$\mathrm{NEP} = \frac{\text{RMS noise current}}{\text{current responsivity}} = \frac{\sigma_i}{\mathcal{R}_i} = \frac{\sigma_i}{\left(\dfrac{i_{\text{signal}}}{\phi_{\text{det}}}\right)} = \frac{\phi_{\text{det}}}{\left(\dfrac{i_{\text{signal}}}{\sigma_i}\right)} = \frac{\phi_{\text{det}}}{\mathrm{SNR}} \tag{8.1}$$

and

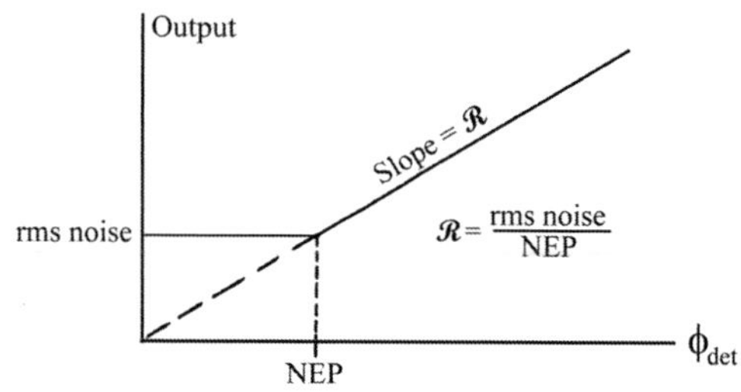

Figure 8.1 Noise-equivalent power.

$$\mathrm{NEP} = \frac{\text{RMS noise voltage}}{\text{voltage responsivity}} = \frac{\sigma_v}{\mathcal{R}_v} = \frac{\sigma_v}{\left(\dfrac{v_{\text{signal}}}{\phi_{\text{det}}}\right)} = \frac{\phi_{\text{det}}}{\left(\dfrac{v_{\text{signal}}}{\sigma_v}\right)} = \frac{\phi_{\text{det}}}{\mathrm{SNR}}. \tag{8.2}$$

NEP is the input flux required to produce an SNR of 1. Thus, the NEP can be considered a minimum detectable flux level. However, an SNR = 1 is only a reference level, and an SNR > 1 (as seen in Fig. 8.2) is usually required for proper system operation. The SNR requirements for the specific application are determined by system-performance metrics such as false-alarm rate, bit-error rate, or image detectability. For example, from Eqs. (8.1) and (8.2), if the SNR requirement is 10, then the input power level required is 10 times the NEP.

There is an analogous quantity for noise-equivalent photon flux (NEPF) [photons/sec]. NEPF is the photon flux required on the detector to produce SNR = 1. For monochromatic flux, we can relate the NEP and NEPF by the usual energy-per-photon quantity $\mathcal{E}$:

$$\mathrm{NEP} = \frac{hc}{\lambda}\mathrm{NEPF}. \tag{8.3}$$

Also defined is the term noise-equivalent irradiance (NEI), which is the irradiance required to produce SNR = 1. Depending on the situation, this irradiance can be specified either at the entrance aperture of the foreoptics in front of the sensor or at the sensor plane itself. The NEI can easily be related

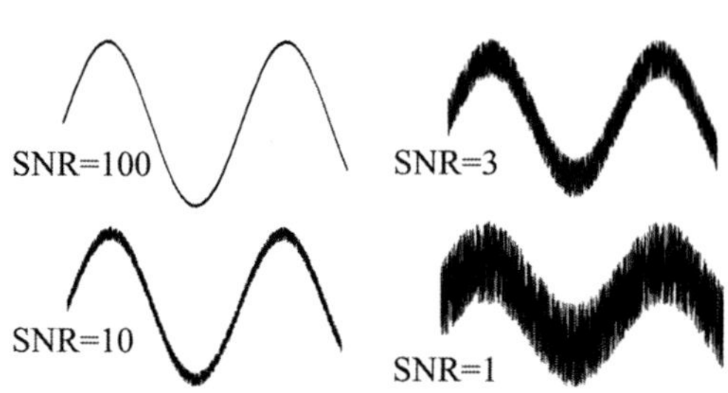

Figure 8.2 Illustration of various SNR levels.

to NEP by simple radiometric conversions. At the aperture plane, NEI × A_{lens} = NEP, assuming that the sensor collects all power in the image plane (the sensor is larger than the spatial extent of the image-plane flux distribution). At the sensor plane, NEI × A_{sensor} = NEP, assuming that the sensor is uniformly illuminated (the sensor is smaller than spatial extent of the image-plane flux distribution).

8.1.1 Numerical-calculation example with NEP

From Eqs. (8.1) and (8.2), NEP is proportional to the rms noise and hence NEP $\propto \sqrt{\Delta f}$ under the usual white-noise assumption. Given the usual dependence of rms noise on the square root of the detector area, NEP $\propto \sqrt{A_{\text{det}}}$ also. We assume that a sensor is measured to have NEP = 0.75 μW, using a bandwidth of 20 kHz. Using this NEP, we can predict the SNR when 1 μW of signal power falls on the detector as SNR = 1 μW/0.75 μW = 1.33. If the bandwidth can be narrowed from 20 kHz to 10 kHz (and still pass the desired signal), the NEP is reduced to 0.54 μW, made smaller (that is, better) by the square root of the bandwidth ratio:

$$\frac{\text{NEP}_1}{\text{NEP}_2} = \frac{\sqrt{\Delta f_1}}{\sqrt{\Delta f_2}}. \tag{8.4}$$

In this case, the system sensitivity has been improved by reducing the measurement bandwidth. The same 1 μW of signal power now gives an SNR of 1 μW/0.54 μW = 1.85.

8.2 Normalized Detectivity *D**

Normalized detectivity *D** is a figure of merit often used in manufacturers' data sheets to specify detector performance. *D** is inversely proportional to NEP. A larger value of *D** indicates better sensitivity. *D** is normalized so that it is independent of the sensor area and measurement bandwidth. Multiplying (1/NEP) by the square root of the detector area and the square root of the measurement bandwidth gives *D** the desired behavior. The definition of *D** is shown in Eq. (8.5):

$$D^* = \frac{\sqrt{A_{\text{det}}}\sqrt{\Delta f}}{\text{NEP}} = \frac{\sqrt{A_{\text{det}}}\sqrt{\Delta f}}{\phi_{\text{det}}}\text{SNR} = \frac{\sqrt{A_{\text{det}}}\sqrt{\Delta f}}{\phi_{\text{det}}}\frac{v_{\text{signal}}}{\sigma_v}. \tag{8.5}$$

Because the rms noise voltage σ_v is generally proportional to $\sqrt{A_{\text{det}}}$ and $\sqrt{\Delta f}$, the dependence of *D** on sensor area and measurement bandwidth cancels out. In any experimental setup for measuring NEP, the sensor area and measurement bandwidth are known. Once the NEP is measured, the multiplication factors are applied to obtain a value for *D**. The benefit of

D^*, from the manufacturers' perspective, is that a value of D^* obtained in this way would apply to a given process-dependent detector technology, regardless of the customer's specific choice of sensor area and measurement bandwidth. This normalization allows for convenient inter-comparison of products from different vendors or comparison of competing sensor technologies, materials, or fabrication processes.

However, to predict the SNR, we must choose the sensor area and noise bandwidth according to the requirements of our application, as discussed in Section 7.3, remembering that a wider bandwidth and larger sensor area lead to more noise. With a radiometric calculation of ϕ_{det} for our measurement configuration, combined with the values for the sensor D^*, sensor area, and measurement bandwidth, the expected SNR can be predicted using

$$\text{SNR} = \frac{\phi_{\text{det}}}{\text{NEP}} = \frac{\phi_{\text{det}} D^*}{\sqrt{A_{\text{det}}}\sqrt{\Delta f}}. \tag{8.6}$$

From Eq. (8.6), D^* has units of [cm $\sqrt{}$Hz/watt]. The typical numerical values are in the range of 10^9 to 10^{12} for most current sensor technologies. Beginning with the definitions of NEP of Eqs. (8.1) and (8.2), we can obtain other expressions for D^*, seen in Eqs. (8.7) and (8.8):

$$D^* = \frac{\sqrt{A_{\text{det}}}\sqrt{\Delta f}}{\text{NEP}} = \frac{\sqrt{A_{\text{det}}}\sqrt{\Delta f}}{\phi_{\text{det}}}\text{SNR} = \frac{\sqrt{A_{\text{det}}}\sqrt{\Delta f}}{\phi_{\text{det}}}\frac{v_{\text{signal}}}{\sigma_v} = \frac{\sqrt{A_{\text{det}}}\sqrt{\Delta f}}{\phi_{\text{det}}}\frac{\mathcal{R}_v}{\sigma_v} \tag{8.7}$$

and

$$D^* = \frac{\sqrt{A_{\text{det}}}\sqrt{\Delta f}}{\text{NEP}} = \frac{\sqrt{A_{\text{det}}}\sqrt{\Delta f}}{\phi_{\text{det}}}\text{SNR} = \frac{\sqrt{A_{\text{det}}}\sqrt{\Delta f}}{\phi_{\text{det}}}\frac{i_{\text{signal}}}{\sigma_i} = \frac{\sqrt{A_{\text{det}}}\sqrt{\Delta f}}{\phi_{\text{det}}}\frac{\mathcal{R}_i}{\sigma_i}. \tag{8.8}$$

From these expressions, we see that D^* is proportional to responsivity, which can be either a spectral or blackbody responsivity. Thus, a spectral $D^*(\lambda)$ or a blackbody $D^*(T)$ can be defined. Because D^* is expressed in terms of watts, it is an energy-based quantity. As such, a plot of the spectral $D^*(\lambda)$ for an ideal photon detector (Fig. 8.3) has the same linear increase with

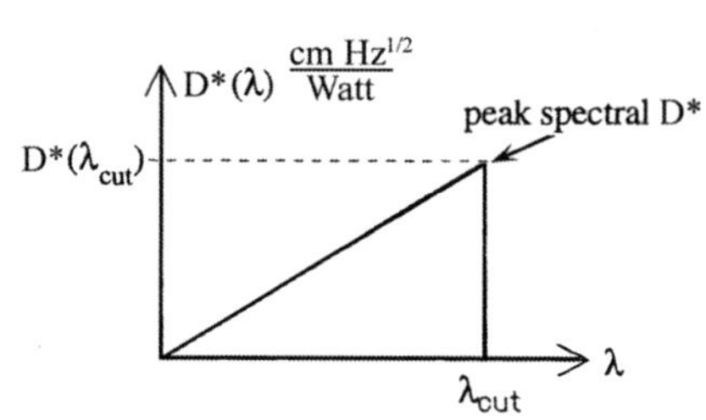

Figure 8.3 Ideal photon-sensor D^*.

wavelength (out to λ_{cutoff}) as seen in Eq. (5.7) and Fig. 5.8 for spectral responsivity. The same *K*-factor of Eq. (5.18) applies to the relationship between peak-spectral $D^*(\lambda_{peak})$ and blackbody $D^*(T)$ as it does for the corresponding responsivity quantities that we discussed in Section 5.2.3.

Validity of the assumptions of the root-area and root-bandwidth dependence of the rms noise are necessary for D^* to be a useful figure of merit. The root-bandwidth dependence is almost always satisfied, since a sensor will not typically be operated over a frequency range that includes significant $1/f$ noise, nor past the inverse of the charge-carrier lifetime. The assumption of a root-area dependence of rms noise requires more careful consideration. This assumption derives from the model of a receiver of a small but finite area, where the power received by the sensor is proportional to its area, with the additional assumption of the sensor noise power being proportional to the cross-sectional area facing the radiation, which we discussed in Section 7.3.

In cases where the root-area dependence on NEP is not met, the relevant figure of merit is the NEP per root bandwidth of a given-sized pixel (where the pixel size is determined by the IFOV requirements). The performance issue is then what SNR is achieved by a sensor of this size. In this context, the relevant normalization of NEP is with respect to root-bandwidth only, yielding units of [W/$\sqrt{}$Hz], with the pixel size specified. This is relevant when immersion lenses or other light concentrators are used in an FPA format, or when antennas are used to couple radiation into a sub-wavelength-sized detector that would otherwise have a very small capture cross section. It is also worth remembering that an SNR calculated from a D^* value is valid only when internally generated sensor noise limits the performance.

8.2.1 Numerical calculations with *D**

D^* is used to choose an appropriate detector technology for a given application. However, NEP remains the fundamental quantity for the calculation of SNR because NEP retains the root-area and root-bandwidth noise dependencies that impact the rms noise:

$$\mathrm{NEP} = \frac{\phi_{det}}{\mathrm{SNR}} = \frac{\sqrt{A_{det}}\sqrt{\Delta f}}{D^*}. \tag{8.9}$$

For example, suppose that we have a detector, specified by $D^* = 3 \times 10^9$ cm Hz$^{1/2}$ W^{-1}. If this sensor has dimensions 1 mm $\times$ 1 mm, and it is used in a 20-kHz bandwidth, then from Eq. (8.9), its NEP will be 4.7 nW. If the same sensor material was made into a 50 μm $\times$ 50 μm detector, and it was used in a bandwidth of 10 kHz, we would find the NEP for this situation (given that D^* is independent of area and bandwidth) to be 0.16 nW. Remember that

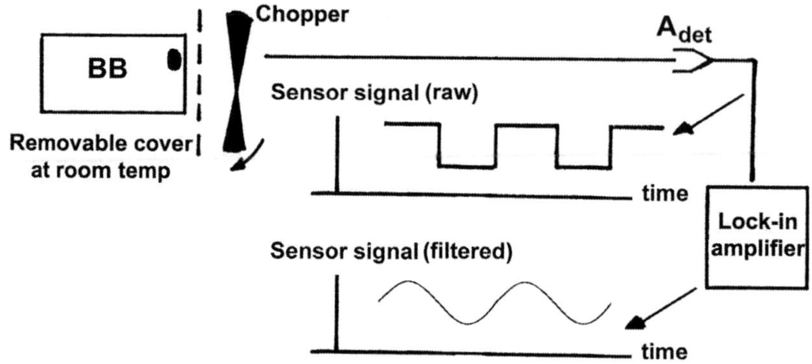

Figure 8.4 D^* measurement configuration.

when the detector area and bandwidth change, NEP changes, but D^* stays constant.

8.2.2 Measurement procedure for D^*

In Fig. 8.4, we see the basic measurement setup for D^*. A blackbody source (typically at 500 K) is configured with a room-temperature chopper wheel such that a known $\Delta\phi_{\text{det}}$ falls on the detector, where $\Delta\phi_{\text{det}} = \phi_{\text{det,BB}} - \phi_{\text{det,chopper}}$. The sensor output voltage v_{signal} is measured within a chosen measurement bandwidth Δf around the fundamental of the chopping frequency. The lock-in amplifier is triggered to the sync signal from the chopper. The voltage responsivity is calculated as $\mathcal{R}_v = v_{\text{signal}}/\Delta\phi_{\text{det}}$. The blackbody source is then blocked by a room-temperature cover, and the rms noise voltage σ_v is measured at the same frequency using the same measurement bandwidth. Using $\text{NEP} = \sigma_v/\mathcal{R}_v$, D^* is calculated from measured data according to Eq. (8.5).

As a sample calculation of D^* from measured data, we assume that the blackbody is at 500 K; with an aperture diameter of 1 cm, the chopper is at 300 K, the detector has an area of 1 mm^2, the distance between the source and detector is 1.5 m, and the chopper modulates the flux at a frequency of 1000 Hz. The measurement bandwidth of the lock-in amplifier is $\Delta f = 10$ Hz. The peak signal voltage is 30 mV, and the rms noise is 30 μV:

$$L = \frac{\sigma}{\pi}(T_{\text{BB}}^4 - T_{\text{chopper}}^4) = \sigma(500^4 - 300^4) = 0.095\,\frac{\text{W}}{\text{cm}^2\ \text{sr}}, \tag{8.10}$$

$$\phi_{\text{det}} = \frac{LA_sA_{\text{det}}}{r^2} = 0.095\frac{\pi}{4}(1)^2\left[\frac{10^{-3}\ \text{m}}{1.5\ \text{m}}\right]^2 = 3.3\times10^{-8}\ \text{W}, \tag{8.11}$$

$$\mathcal{R}_v = 30\ \text{mV}/3.3\times10^{-8}\ \text{W} = 9.09\times10^5\ \text{V/W}, \tag{8.12}$$

$$\text{NEP} = \frac{30\ \mu\text{V}}{9.09 \times 10^5\ \text{V/W}} = 3.3 \times 10^{-11}\ \text{W}, \tag{8.13}$$

and

$$D^* = \frac{0.1\ \text{cm}\ 3.16\sqrt{\text{Hz}}}{3.3 \times 10^{-11}\ \text{W}} = 9.6 \times 10^9\ \text{cm}\ \sqrt{\text{Hz}}/\text{W}. \tag{8.14}$$

8.3 Background-Limited *D** for Photon Sensors

The best SNR for IR imagers that use photon sensors is usually obtained with background-limited (so-called *BLIP*) performance, unless the imager is in a very low-background situation such as a cold-space background. As such, it is important to understand the implications of this limit and how those limitations are communicated in manufacturers' spec sheets. The expression we will develop for D^*_{BLIP} will be useful for our systems discussions in Chapter 9. It is intuitive that D^*_{BLIP} will depend on the solid angle that the background subtends as seen from the sensor as well as on the temperature of the background. A high image-space *F/#* is a narrow cone angle, which will lead to less background flux on the detector (and thus less shot noise on the background), and hence a higher D^*_{BLIP}.

We begin our analysis with a monochromatic signal and background fluxes $\phi_{q,\text{sig}}$ and $\phi_{q,\text{bkg}}$ [photons/sec]. The current from signal flux is

$$i_{\text{sig}} = \eta q \phi_{q,\text{sig}}, \tag{8.15}$$

and the current from background flux is

$$i_{\text{bkg}} = \eta q \phi_{q,\text{bkg}}. \tag{8.16}$$

The condition for BLIP operation is $\bar{i} = i_{\text{sig}} + i_{\text{bkg}} \approx i_{\text{bkg}}$. The rms shot noise on the background-generated photocurrent is

$$\sigma_i = \sqrt{2qi_{\text{bkg}}\Delta f} = \sqrt{2q(\eta q \phi_{q,\text{bkg}})\Delta f}. \tag{8.17}$$

The SNR for this situation is

$$\text{SNR} = \frac{\eta q \phi_{q,\text{sig}}}{\sqrt{2q(\eta q \phi_{q,\text{bkg}})\Delta f}} = \frac{\sqrt{\eta}\phi_{q,\text{sig}}}{\sqrt{2\Delta f}\sqrt{\phi_{q,\text{bkg}}}}. \tag{8.18}$$

We then set SNR = 1 and solve for the resulting $\phi_{q,\text{sig}}$, which is the noise-equivalent photon flux (NEPF) in the background-noise limit:

$$\mathrm{NEPF}_{\mathrm{BLIP}} = \frac{\sqrt{2\Delta f}\sqrt{\phi_{q,\mathrm{bkg}}}}{\sqrt{\eta}}. \tag{8.19}$$

The units are consistent: [Hz$^{1/2}$ (1/sec)$^{1/2}$] = [photons/sec]. Under the assumption of monochromatic flux of wavelength λ, changing NEPF [photons/sec] to NEP [W] lets us obtain an expression for the background-limited NEP,

$$\mathrm{NEP}_{\mathrm{BLIP}} = \frac{hc}{\lambda}\mathrm{NEPF}_{\mathrm{BLIP}} = \frac{hc}{\lambda}\frac{\sqrt{2\Delta f}\sqrt{\phi_{q,\mathrm{bkg}}}}{\sqrt{\eta}} = \frac{hc}{\lambda}\frac{\sqrt{2\Delta f}\sqrt{A_{\mathrm{det}} \times E_{q,\mathrm{bkg}}}}{\sqrt{\eta}}, \tag{8.20}$$

where we have written the background photon flux $\phi_{q,\mathrm{bkg}}$ on the detector in terms of the background photon irradiance $E_{q,\mathrm{bkg}}$ and sensor area. With this NEP, we can write an expression for D^*_{BLIP}, still monochromatic at wavelength λ:

$$D^*_{\mathrm{BLIP}}(\lambda) = \frac{\sqrt{A_{\mathrm{det}}}\sqrt{\Delta f}}{\mathrm{NEP}_{\mathrm{BLIP}}} = \frac{\lambda}{hc}\frac{\sqrt{\eta}}{\sqrt{2}}\frac{1}{\sqrt{E_{q,\mathrm{bkg}}}}. \tag{8.21}$$

The units are consistent: [J^{-1} × (sec cm^2)$^{1/2}$] = [cm Hz$^{1/2}$ W^{-1}]. Now, we let the background flux be distributed over λ, as from a blackbody at temperature T. The only background photons that "count" toward the shot noise are those that have $\lambda < \lambda_{\mathrm{cut}}$, so we can express the integrated background photon irradiance as

$$E_{q,\mathrm{bkg}} \equiv \int_0^{\lambda_{\mathrm{cut}}} E_{q,\mathrm{bkg}}(\lambda, T_{\mathrm{bkg}})\mathrm{d}\lambda, \tag{8.22}$$

which yields an expression for $D^*_{\mathrm{BLIP}}(\lambda, T_{\mathrm{bkg}})$:

$$D^*_{\mathrm{BLIP}}(\lambda, T_{\mathrm{bkg}}) = \frac{\sqrt{A_{\mathrm{det}}}\sqrt{\Delta f}}{\mathrm{NEP}} = \frac{\lambda}{hc}\frac{\sqrt{\eta}}{\sqrt{2}}\frac{1}{\sqrt{\int_0^{\lambda_{\mathrm{cut}}} E_{q,\mathrm{bkg}}(\lambda, T_{\mathrm{bkg}})\mathrm{d}\lambda}}. \tag{8.23}$$

The quantity in Eq. (8.23) is still a spectral D^*, that is, an explicit function of wavelength with the usual linear dependence of D^* on λ for a photon detector (valid up to the cutoff wavelength).

What is typically plotted in manufacturers' spec sheets (as seen in the dashed curves of Fig. 8.5) is the peak-spectral D^* as a function of the cutoff wavelength: $D^*_{\mathrm{BLIP}}(\lambda_{\mathrm{cut}})$. This is the peak value of the $D^*_{\mathrm{BLIP}}(\lambda)$ curve and represents the best possible performance given a certain level of background-photon irradiance. Assuming that $\eta = 1$, this D^*_{BLIP} is plotted as a function of the cutoff wavelength λ_{cut} for a given background temperature. This D^* is the

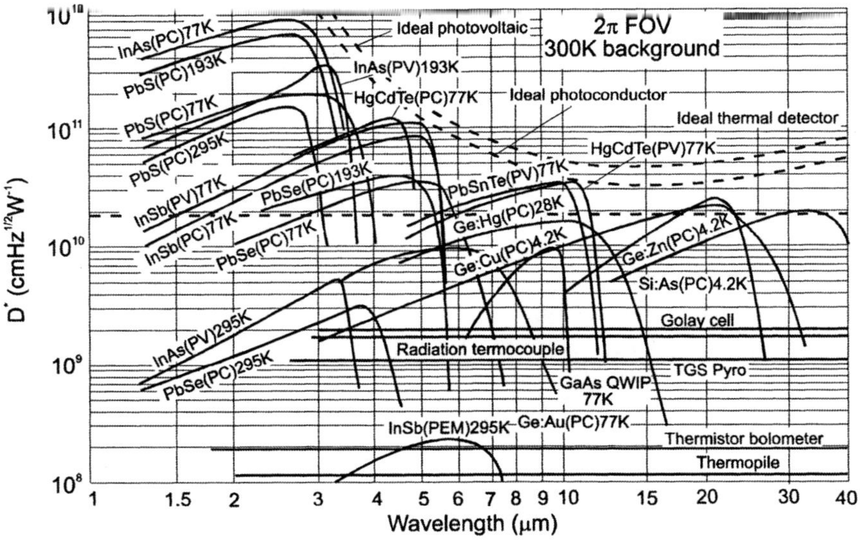

Figure 8.5 Typical manufacturer's specification sheet for D^* (adapted from Ref. 1).

best that can be achieved for a background-limited situation. Substituting λ_{cut} for λ in Eq. (8.23), we find that

$$D^*_{BLIP}(\lambda_{cut}, T_{bkg}) = \frac{\lambda_{cut}}{hc}\frac{\sqrt{\eta}}{\sqrt{2}}\frac{1}{\sqrt{\int_0^{\lambda_{cut}} E_{q,bkg}(\lambda, T_{bkg})d\lambda}}. \tag{8.24}$$

For a hemispherical background (Section 3.5.5), we take $E_{q,bkg} = M_{q,bkg}$:

$$D^*_{BLIP}(\lambda_{cut}, T_{bkg}) = \frac{\lambda_{cut}}{hc}\frac{\sqrt{\eta}}{\sqrt{2}}\frac{1}{\sqrt{\int_0^{\lambda_{cut}} M_{q,bkg}(\lambda, T_{bkg})d\lambda}}. \tag{8.25}$$

We use the usual blackbody expression for radiant exitance and convert to photon exitance:

$$D^*_{BLIP}(\lambda_{cut}, T_{bkg}) = \frac{\lambda_{cut}}{hc}\frac{\sqrt{\eta}}{\sqrt{2}}\frac{1}{\sqrt{\int_0^{\lambda_{cut}} M_{bkg}(\lambda, T)\frac{\lambda}{hc}d\lambda}}, \tag{8.26}$$

$$D^*_{BLIP}(\lambda_{cut}, T_{bkg}) = \frac{\lambda_{cut}}{hc}\frac{\sqrt{\eta}}{\sqrt{2}}\frac{1}{\sqrt{\int_0^{\lambda_{cut}} \frac{2\pi hc^2}{\lambda^5[e^{hc/\lambda kT_{bkg}} - 1]}\frac{\lambda}{hc}d\lambda}}, \tag{8.27}$$

and

$$D^*_{\mathrm{BLIP}}(\lambda_{\mathrm{cut}}, T_{\mathrm{bkg}}) = \frac{\lambda_{\mathrm{cut}}}{hc}\frac{\sqrt{\eta}}{\sqrt{2}}\frac{1}{\sqrt{\int_0^{\lambda_{\mathrm{cut}}} \frac{2\pi c}{\lambda^4\left[e^{hc/\lambda k T_{\mathrm{bkg}}} - 1\right]} d\lambda}}. \tag{8.28}$$

Equation (8.28) is the peak-spectral D^* vs cutoff wavelength, parameterized on background temperature, assuming a background-limited photovoltaic sensor. This is plotted as the upper dashed curve in Fig. 8.5, assuming unit quantum efficiency. The assumption is that the shot noise on the background-generated photoelectrons is the limiting noise, which is the case of BLIP photovoltaic sensors. If the limiting noise is generation–recombination (G-R) noise on the background, we find a $\sqrt{2}$ higher rms noise (Section 7.9), and the best-possible D^* curve is pulled down by a factor of $1/\sqrt{2}$ from the photovoltaic case. The lower dashed curve of Fig. 8.5 applies to BLIP photoconductors.

The solid lines in Fig. 8.5 are the $D^*(\lambda)$ curves for various specific detectors. For any detector, as the cutoff wavelength is approached, the ideal $D^*(\lambda)$ curve increases linearly with wavelength, consistent with the spectral responsivity for a photon sensor plotted in energy-based units, seen in Fig. 5.8. The spectral D^* curves for specific detectors generally follow this behavior because $D^*(\lambda)$ is proportional to spectral responsivity $\mathcal{R}(\lambda)$ as seen in Eqs. (8.7) and (8.8). Note that Fig. 8.5 is plotted on a log scale for both axes.

For the dashed-line plots in Fig. 8.5, the quantum efficiency η is assumed to be 1. If $\eta \neq 1$, D^* is proportional to $\sqrt{\eta}$ as seen in Eq. (8.28). If the (solid-line) $D^*(\lambda)$ curve for a given BLIP photon detector does not touch the (dashed-line) peak-spectral D^* curve at the cutoff wavelength of the detector, we have evidence of non-unit quantum efficiency.

If the photon sensor is not operating in the background limit, other sources of noise contribute, and the D^* will not be as high as for the background-limited cases in Fig. 8.5. In this context, the $D^*_{\mathrm{BLIP}}(\lambda_{\mathrm{cutoff}}, T_{\mathrm{bkg}})$ of Eq. (8.28) represents the best possible D^* for a sensor with a given cutoff wavelength.

The plots in Fig. 8.5 assume a hemispherical blackbody background temperature of 300 K. In Fig. 8.6, we plot $D^*_{\mathrm{BLIP}}(\lambda_{\mathrm{cutoff}}, T_{\mathrm{bkg}})$ for other temperatures, assuming an ideal photovoltaic sensor. Colder background temperatures yield less background flux, which leads to higher background-limited D^*s. Also, for colder backgrounds, the peak of the background spectral photon exitance shifts to longer wavelengths according to Wien's law for photon flux discussed in Section 4.3.3. Thus as T_{bkg} decreases, the $D^*_{\mathrm{BLIP}}(\lambda_{\mathrm{cutoff}}, T_{\mathrm{bkg}})$ curves themselves move upward, while the minima of these curves shift toward longer wavelengths.

Usually, in practice, an IR sensor views the background through a limiting aperture, either a cold shield or cold stop inside a cryogenic dewar, as seen in Fig. 7.12, or the exit pupil of the foreoptics. This reduces the

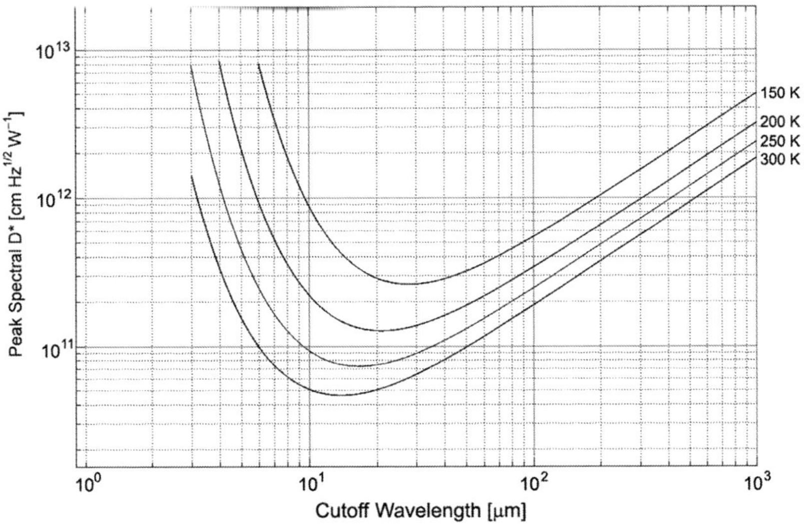

Figure 8.6 $D^*_{BLIP}(\lambda_{cutoff})$ for hemispherical background temperatures of 150, 200, 250 and 300 K, assuming an ideal photovoltaic sensor.

background irradiance at the detector as compared to the hemispherical-background cases just considered, increasing the sensor's D^* under the assumption that the sensor continues to operate in the background limit. To develop the dependence of D^*_{BLIP} on the *F*/# subtended by the background, we begin with Eq. (8.24). The irradiance at the detector is a function of the *F*/# of the cone of radiation reaching the sensor and of the background radiance L. We previously considered this situation in Eq. (3.26) in the context of irradiance described in energy-based units. The relation for large *F*/#s is $E_{\text{image}} \approx \pi L_{\text{obj}}/4(F/\#)^2_{\text{image-space}}$, which also applies between photon radiance and photon irradiance. Substituting Eq. (3.26) into Eq. (8.24), we find that

$$D^*_{\text{BLIP}}(\lambda_{\text{cut}}, T_{\text{bkg}}) \approx \frac{\lambda_{\text{cut}}}{hc}\sqrt{\frac{2\eta}{\pi}}\frac{F/\#}{\sqrt{\int_0^{\lambda_{\text{cut}}} L_{q,\text{bkg}}(\lambda, T_{\text{bkg}})d\lambda}}. \tag{8.29}$$

As expected, D^* increases for increasing *F*/#, assuming that the sensor continues to operate under background-limited conditions. However, as the *F*/# increases, the background contribution will decrease to the point where some other noise source will dominate.

8.4 Johnson-Noise-Limited (JOLI) *D** for Photon Sensors

Johnson-noise-limited (JOLI) photon-sensor operation is obtained at very low backgrounds, so that shot noise on the background can be neglected. From Eq. (7.24), the Johnson-noise-limited SNR is

$$\mathrm{SNR} = \frac{\phi_{q,\mathrm{sig}}\eta q}{\sqrt{\frac{4kT}{R_{\mathrm{det}}}\Delta f}}. \tag{8.30}$$

We take the usual approach of setting SNR = 1 and solving for $\phi_{q,\mathrm{sig}}$ as the NEPF:

$$\mathrm{NEPF} = \frac{1}{\eta q}\sqrt{\frac{4kT\Delta f}{R_{\mathrm{det}}}}. \tag{8.31}$$

We then convert to NEP, assuming monochromatic signal flux:

$$\mathrm{NEP} = \frac{hc}{\lambda\eta q}\sqrt{\frac{4kT\Delta f}{R_{\mathrm{det}}}}. \tag{8.32}$$

Using Eq. (8.5), we find that

$$D^*(\lambda) = \frac{\lambda\eta q}{2hc}\sqrt{\frac{R_{\mathrm{det}}A_{\mathrm{det}}}{kT}}. \tag{8.33}$$

The Johnson-noise-limited D^* is seen to be proportional to the square root of the *RA product*, which is a figure of merit for materials intended for low-background photon sensors. The significance of the *RA* product can be understood with reference to the plot of D^* vs background irradiance (Fig. 8.7).

For high background irradiance, the sensor performance is background limited. As the background irradiance is reduced, the shot noise on the background decreases in proportion to the square root of the photon irradiance (shown as the straight line of slope −1/2 on a log–log plot), increasing D^* while the sensor continues to operate in the background limit. When the background shot noise becomes negligible compared to the Johnson

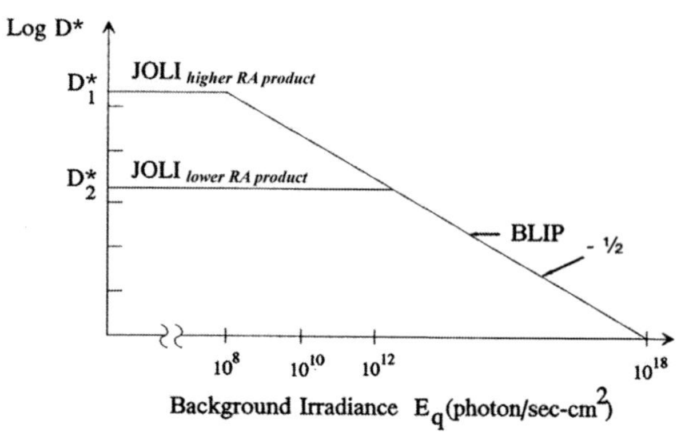

Figure 8.7 D^* vs background irradiance on the detector.

noise, D^* reaches a plateau. Beyond this, we find no further increase in D^* for further decreases in background irradiance because the sensor is operating in the Johnson-noise limit. Sensors with higher RA products are desirable because their transition from background-limited performance to Johnson-noise-limited performance occurs at smaller values of background irradiance, allowing higher D^*s to be achieved when the background irradiance is negligible.

8.5 Temperature-Fluctuation-Noise-Limited D^* for Thermal Sensors

We begin with Eq. (7.31) for the SNR of a thermal-fluctuation-limited thermal sensor:

$$\mathrm{SNR} = \frac{\phi_{\mathrm{incident}}\sqrt{\varepsilon_{\mathrm{det}}}}{\sqrt{8A_{\mathrm{det}}\sigma k(T_{\mathrm{det}}^5 + T_{\mathrm{bkg}}^5)\Delta f}}. \tag{8.34}$$

Following the usual procedure, we set SNR = 1 and solve for ϕ_{incident} to yield the NEP, which has the expected root-area and root-bandwidth dependence

$$\mathrm{NEP} = \frac{\sqrt{8A_{\mathrm{det}}\sigma k(T_{\mathrm{det}}^5 + T_{\mathrm{bkg}}^5)\Delta f}}{\sqrt{\varepsilon_{\mathrm{det}}}}. \tag{8.35}$$

Converting NEP to D^*, we find that

$$D^* = \frac{\sqrt{A_{\mathrm{det}}}\sqrt{\Delta f}}{\mathrm{NEP}} = \frac{\sqrt{\varepsilon_{\mathrm{det}}}}{\sqrt{8\sigma k(T_{\mathrm{det}}^5 + T_{\mathrm{bkg}}^5)}}. \tag{8.36}$$

The ideal thermal-sensor spectral responsivity, and hence spectral D^*, is flat with wavelength, limited only by the spectral absorption of the sensor material itself:

$$D^*(\lambda) = \frac{\sqrt{A_{\mathrm{det}}}\sqrt{\Delta f}}{\mathrm{NEP}} = \frac{\sqrt{\varepsilon_{\mathrm{det}}(\lambda)}}{\sqrt{8\sigma k(T_{\mathrm{det}}^5 + T_{\mathrm{bkg}}^5)}}. \tag{8.37}$$

For a unit-emissivity sensor, when both the sensor and the background are at 300 K, we find that $D^* = 1.9 \times 10^{10}$ cm Hz$^{1/2}$/W, which is the upper limit to D^* for an uncooled sensor looking into a room-temperature background. If the sensor is cooled to 77 K, we obtain only about a factor of $\sqrt{2}$ improvement to $D^* = 2.7 \times 10^{10}$ cm Hz$^{1/2}$/W because of the sum in the denominator of Eq. (8.37) and the 5th-power dependence on temperature. In the case of a room-temperature background, further cooling of the sensor

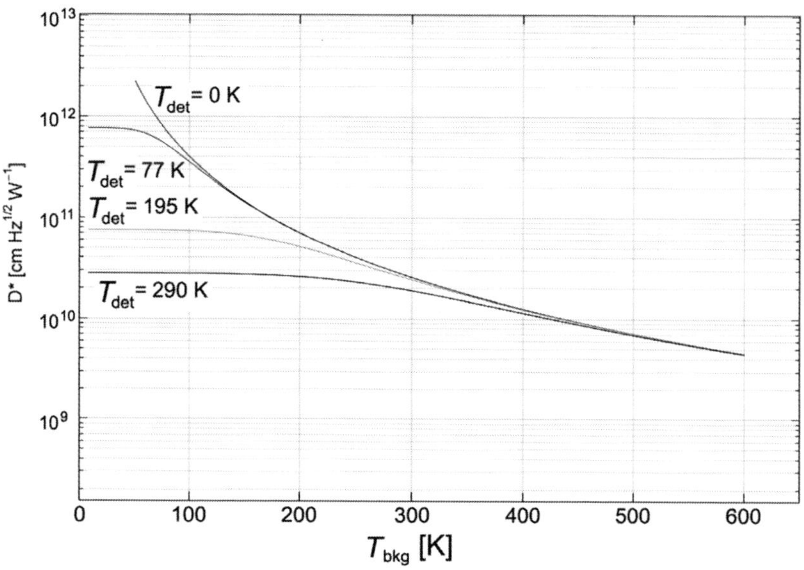

Figure 8.8 *D** for a thermal-fluctuation-limited thermal detector as a function of both detector and background temperatures.

alone produces no appreciable improvement in D^*. However, if both the background and the sensor are cold, substantial gains in thermal-fluctuation-limited D^* can be achieved, as seen in the plot of Fig. 8.8, which is why thermal sensors are cooled in low-background applications such as astronomy. Note that current commercially available uncooled camera systems are not yet operating in the thermal-fluctuation-noise limit.

8.6 Comparing the Performance of Thermal and Photon Sensors

We can use the results of Sections 8.3 and 8.5 to compare the ultimate possible performance from ideal thermal sensors and ideal photon sensors. Assuming a 300 K background (a ground-based system, or a down-looking system with the Earth in the FOV), a BLIP photovoltaic cooled sensor with a 10-μm cutoff wavelength can achieve a peak spectral D^* of 5×10^{10} cm Hz$^{1/2}$ W^{-1}. For this photon sensor, D^*s at shorter wavelengths will decrease linearly with wavelength, as seen in Fig. 8.3. Under the same conditions, an uncooled thermal sensor assumed to be operating in the thermal-fluctuation limit could achieve a D^* of 1.9×10^{10} cm Hz$^{1/2}$ W^{-1}, independent of wavelength. With cooling to 77 K, this thermal sensor could achieve a D^* of 2.7×10^{10} cm Hz$^{1/2}$ W^{-1}, also independent of wavelength. We compare these two situations in Fig. 8.9.

For operating wavelengths in the MWIR and LWIR, uncooled thermal-sensor camera systems have generally lower SNR performance than cooled

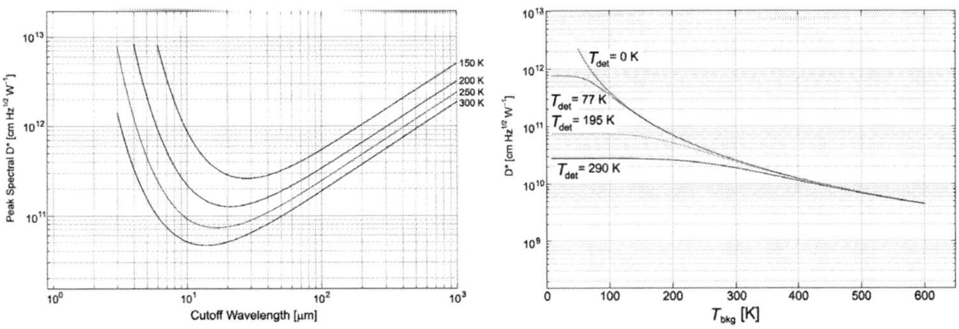

Figure 8.9 Comparing D^* for (left) an ideal BLIP photovoltaic sensor to (right) an ideal temperature-fluctuation-limited thermal sensor.

photon-sensor camera systems. However, thermal sensors do not require cooling and can be used to detect at wavelengths for which photon sensors would need to be cooled. At wavelengths in the far-infrared and longer, photon detectors become less desirable because of the stringent cooling (significantly lower than 77 K) required for operation with the very small-bandgap semiconductors, consistent with long-wavelength response, as discussed in Section 5.3. In this longwave region, thermal detectors become the sensors of choice, with excellent D^* values possible in low-background situations.

Reference

1. A. Rogalski, "Infrared thermal detectors versus photon detectors: I. Pixel performance," *Proc. SPIE* **3182**, 14–25 (1997) [doi: 10.1117/12.280417].

Bibliography

Dereniak, E. L., and G. D. Boreman, *Infrared Detectors and Systems*, Wiley, New York, (1996).

Vincent, J. D., S. E. Hodges, J. Vampola, M. Stegall, and G. Pierce, *Fundamentals of Infrared and Visible Detector Operation and Testing, 2nd Edition*, Wiley, New York (2016).

Chapter 9
Infrared Systems

9.1 General Comments

In this chapter, we will develop design equations for IR systems that relate the optics parameters and the sensor D^* to the signal-to-noise ratio (SNR) achieved by the system. We distinguish between an *IR search system*, which is designed to detect and locate a point source within a specified search volume, and a *thermal-imager system*, which is designed to yield a map of temperature and emissivity of extended-source scenes.

We first consider the relationship of the scan format to the signal bandwidth required, which will determine the SNR. We then develop the range equation for a search system design. Following that, we develop expressions for the noise-equivalent temperature difference (NETD) for a thermal-imager system. We then introduce the minimum resolvable temperature difference (MRTD) as a performance descriptor that combines thermal sensitivity and spatial resolution.

9.2 Scan Formats

The three-dimensional distribution of object-space brightness is mapped by the optical system onto a two-dimensional image plane, according to the principles outlined in Chapter 2. The image plane is populated with sensors whose dimensions, along with the focal length of the optics, determine the spatial density of pixels projected onto the object scene. The instantaneous field of view (IFOV) is that portion of object space being looked at by a given detector at one instant in time. The IFOV is the footprint of the detector, which can be either stationary or moving as a function of time. We define horizontal FOV (HFOV), vertical FOV (VFOV), horizontal IFOV (HIFOV), and vertical IFOV (VIFOV), as shown in Fig. 9.1.

Scanning can be used to cover a two-dimensional FOV with a number of detectors that is less than the number of pixels. We begin our discussion with a single detector scanned in two orthogonal directions. In IR systems, this serial-scan format is no longer used, but the associated analysis serves to

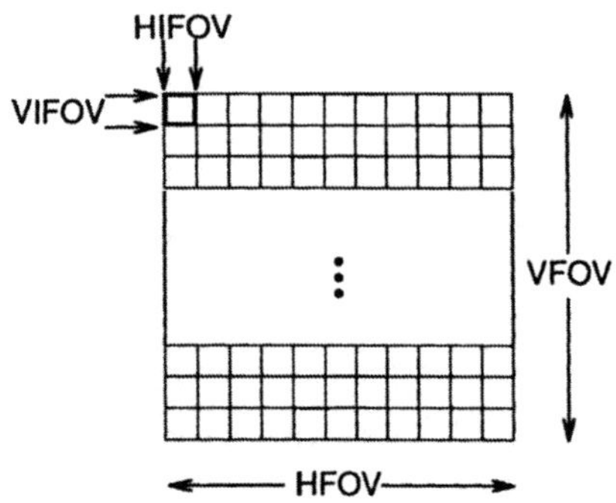

Figure 9.1 Definition of horizontal and vertical fields of view.

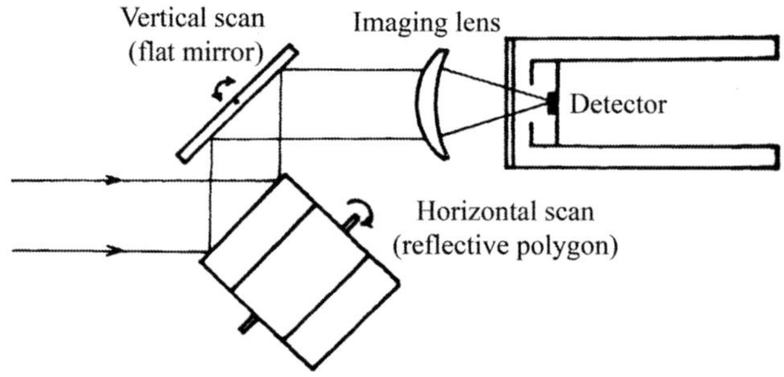

Figure 9.2 Scanning an IFOV across a two-dimensional scene (adapted from Ref. 1).

highlight the bandwidth dependence on scan format. This scan format is occasionally used for systems in developmental parts of the spectrum, such as the far infrared or terahertz. The scanning of the optical line of sight is accomplished with mirrors, as shown in Fig. 9.2. Rotating wedge prisms can also be used to move the line of sight.

The resulting IFOV scan is shown in Fig. 9.3. The IFOV is scanned horizontally along one line of the image, then stepped down a line and scanned again, repeating until data for the whole VFOV is obtained. The detector output is read out continuously as the IFOV moves.

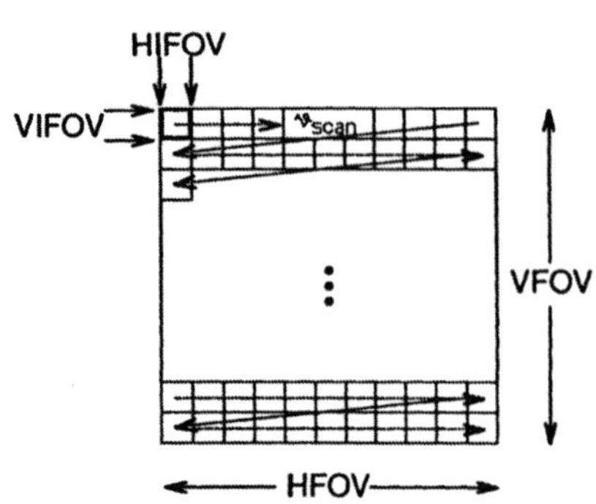

Figure 9.3 Serial scan of single detector.

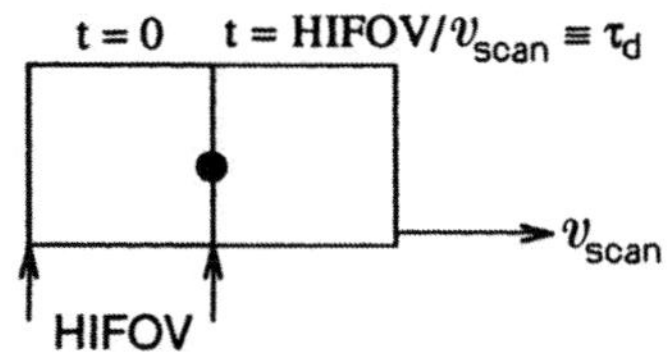

Figure 9.4 Definition of dwell time.

Considering the scan velocity v_{scan}, the moving detector footprint remains on a given image feature (denoted as the dark spot in Fig. 9.4) for a dwell time of

$$\tau_d = \mathrm{HIFOV}/v_{scan}. \tag{9.1}$$

For a single-detector system, the dwell time is the frame time divided by the number of pixels: $\tau_d = T_{frame}/N_{pixels}$. This expression assumes unit scan efficiency; that is, we neglect such issues as overscan and fly-back time that will increase the required scan velocity and reduce the dwell time for a given fixed frame time. In this arrangement, the frame time is the number of lines in the image multiplied by the time it takes to scan each line:

$$T_{frame} = N_{lines} \times T_{line} = \frac{\mathrm{VFOV}}{\mathrm{VIFOV}} \times \frac{\mathrm{HFOV}}{v_{scan}}. \tag{9.2}$$

To pass the scene data, a scanned system requires a bandwidth of $\Delta f = 1/(2\tau_d)$. A scanning system covers the whole FOV with one detector, and it must take all the samples in one frame time. A slower scan velocity reduces the required bandwidth and is better for averaging out random noise. The tradeoff is that it takes longer to scan the whole scene when we scan more slowly.

A staring system has a dedicated sensor for each image pixel. A staring array looks at the whole FOV continously with a dwell time equal to the frame time: $\tau_d = T_{frame}$. It operates at a narrower bandwidth by a factor of $1/N_{sensors}$ compared to a single-sensor serial scan system. Comparing SNR for a scanned single-detector system vs a staring system with same T_{frame}, the staring system has a lower rms noise and hence a better SNR by a factor of $\sqrt{N_{sensors}}$ for a given level of scene flux:

$$\frac{\mathrm{SNR}_{staring}}{\mathrm{SNR}_{scanned}} = \sqrt{\frac{\Delta f_{scanned}}{\Delta f_{staring}}} = \sqrt{\frac{T_{frame}}{\tau_d}} = \sqrt{N_{sensors}}. \tag{9.3}$$

Equation (9.3) is a general result. The SNR advantage of adding sensors goes as the square root of the number of sensors used. This applies to multiple-detector scan formats as well, which we consider next.

9.2.1 Multiple-detector scan formats

Even if there are not as many detectors as pixels in the image, additional sensors can still be used to give the SNR advantage of Eq. (9.3). The detectors can be arranged in various configurations, and the same SNR advantage is achieved for each.

A parallel scan uses multiple sensors in the cross-scan direction, as seen in Fig. 9.5. For a given fixed frame time, we can use a slower scan velocity because multiple horizontal lines are covered at once, resulting in a longer dwell time and a narrower noise bandwidth. The scan format seen in Fig. 9.5 requires a horizontal and a vertical scan mirror. If there are enough sensors in a one-dimensional array to cover the whole VFOV simultaneously, then only a horizontal scan mirror is needed, which is termed a *pushbroom scan*. The pushbroom-scan motion can also be accomplished by a linear motion of the whole camera platform, for instance with a downlooking sensor mounted on an airplane.

A serial-scan system shown in Fig. 9.6 uses multiple sensors in the along-scan direction. The signal from each sensor is collected and co-added with the signal from other sensors as the scan progresses. This is called *time delay and integration* (TDI). The signal goes up by N, and the (uncorrelated) noise from

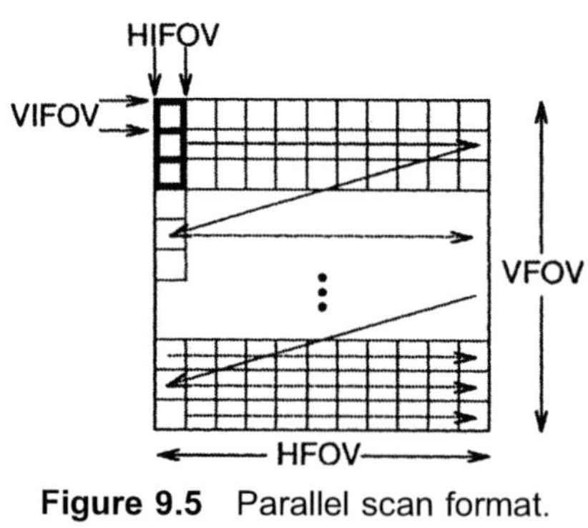

Figure 9.5 Parallel scan format.

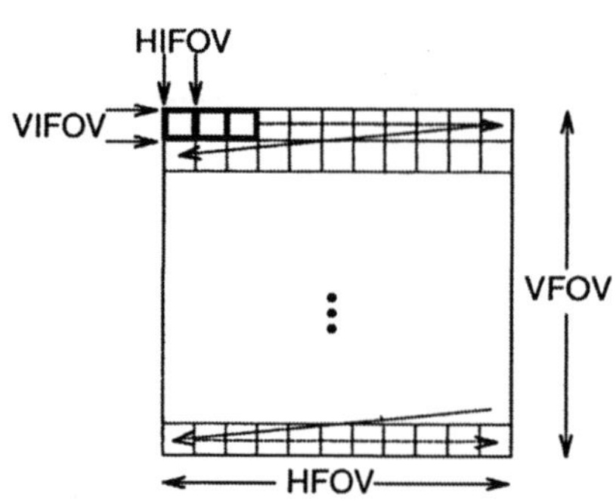

Figure 9.6 Serial scan format.

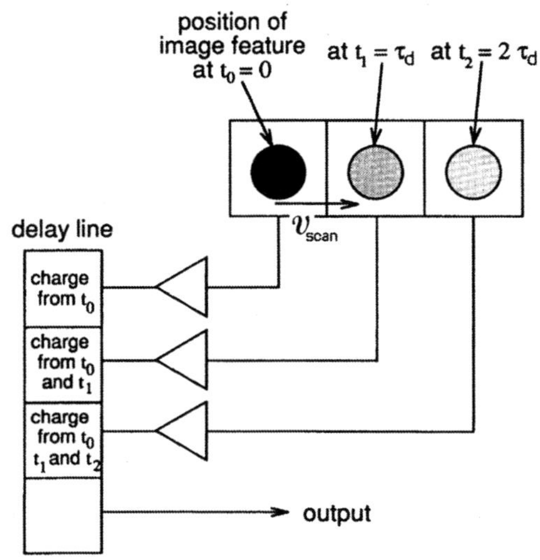

Figure 9.7 Delay line required to implement time delay and integration.

the N measurements adds together in quadrature, so the noise goes up by $\sqrt{N}$. Thus, the serial-scan format has $\sqrt{N_{sensors}}$ better SNR compared to a single-detector scan, equivalent to having a detector with N times the HIFOV (hence a factor of N more dwell time), but the TDI keeps the spatial resolution of the original sensor size. A disadvantage to TDI is that it requires a delay line (Fig. 9.7), typically a charge-coupled device, to move the signal charge synchronized with the horizontal-scan motion of the IFOVs.

The time-delay-and-integration procedure can be best undersood by imagining that the scene scans across a fixed detector array. First, signal charge is accumulated from sensor #1 for a time τ_d. Then, the scene scans one pixel in the horizontal direction, accumulating signal charge from sensor #2. The stored charge in the delay line from sensor #1 is added to the charge from sensor #2. This operation is repeated for each sensor in TDI. Again, the signal grows proportionally to $N_{sensors}$, and rms noise grows as $\sqrt{N_{sensors}}$, with SNR being proportional to $\sqrt{N_{sensors}}$.

9.3 Search System: Range Equation

The end product of a search system is not an image, but rather a decision as to whether or not a point source of given intensity is present within the FOV of the sensor. The relevant receiver operating characteristics (such as the probability of correct detection or false-alarm rate) are improved by increasing the SNR. In the context of a search system, an object of interest will first be noticed when it is far away and acts as a point source. The sensitivity of modern IR sensor systems is such that an object can be detected as a point target before it is close enough to be resolved by the optics. As noted in Sections 2.6 and 2.7.1, the smallest object feature that can be resolved is the distance to the object multiplied by either the diffraction-limited angular resolution of 2.4 λ/D or the image-space angular subtense of a detector $\sqrt{A_{det}}/f$. The larger of these two object-plane footprints will determine when a given object transitions from point-source behavior to becoming resolvable. The design equation for search system design is expressed in terms of the distance at which a point source of radiation can be reliably detected by a system with a given set of specifications. Our treatment assumes unit transmittance for both the atmosphere and optical elements. Atmospheric attenuation[2,3] over the chosen spectral passband will reduce the detection range below the ideal values presented.

We will investigate the tradeoffs involved in the design of an IR search system as seen below. We have a choice of sensor D^*, frame rate, optics configuration, and scan format. The general design objective is to be able to detect the point source, specified by its intensity in watts/steradian, as far away as possible. With reference to the situation shown in Fig. 9.8, the flux collected by the system is $\phi_{det} = (I \times A_{optic})/R^2$, assuming that the detector captures all of the flux transmitted by the lens.

We can express the signal voltage as

$$v_{sig} = \mathcal{R}_v \times \phi_{det} = \mathcal{R}_v \times \frac{A_{optic} \times I}{R^2}. \tag{9.4}$$

Dividing both sides by the rms noise voltage σ_v yields the SNR:

$$\text{SNR} = \frac{v_{sig}}{\sigma_v} = \frac{\mathcal{R}_v}{\sigma_v} \times \frac{A_{optic} \times I}{R^2}. \tag{9.5}$$

Recalling Eq. (8.5), we can express the SNR as

$$\text{SNR} = \frac{1}{\text{NEP}} \times \frac{A_{optic} \times I}{R^2} = \frac{D^*}{\sqrt{A_{det}\Delta f}} \times \frac{A_{optic} \times I}{R^2}. \tag{9.6}$$

The interpretation of Eq. (9.6) is that the SNR produced by the system increases as optic aperture $A_{optic}\uparrow$, point-source intensity $I\uparrow$, and sensor

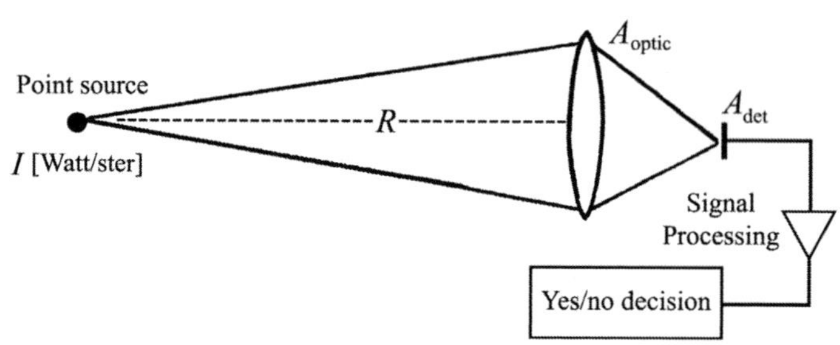

Figure 9.8 Search system configuration.

detectivity $D^*\uparrow$. The SNR decreases as range $R\uparrow$, noise bandwidth $\sqrt{\Delta f}\uparrow$, and sensor area $\sqrt{A_{det}}\uparrow$. Remember that we are collecting radiation from a point source, not an extended source, and we assume that the sensor dimension is chosen large enough to collect the image-irradiance distribution of the point source. In this situation, making A_{det} larger does not increase the signal but does increase the internal detector noise. We assume in our development that the sensor is not background limited so that the D^* is independent of the *F*/#.

We can now solve Eq. (9.6) for the maximum range at which the point source can be detected:

$$R = \left[\frac{A_{optic} I D^*}{\text{SNR}\sqrt{\Delta f A_{det}}}\right]^{1/2}. \tag{9.7}$$

In Eq. (9.7), the SNR is interpreted as that required to reach a decision with an acceptable degree of certainty. If, as shown in Fig. 9.9, a higher SNR is required to give better certainty of a result, then the search system's range is reduced—it cannot see as far.

We can express the standoff range in terms of the optics $F/\# = f/D$ and the pixel solid-angle subtense $\Omega = A_{det}/f^2$. With these substitutions, Eq. (9.7) becomes

$$R = \left[\frac{(\pi/4)D^2\, I\, D^*}{\text{SNR}\sqrt{\Delta f A_{det}}}\right]^{1/2} = \left[\frac{(\pi/4)D^2\, I\, D^*}{\text{SNR}\sqrt{\Delta f}\sqrt{\Omega} f}\right]^{1/2} = \left[\frac{I\, D^*}{\text{SNR}\sqrt{\Delta f}}\frac{\pi/4\, D}{F/\#\ \sqrt{\Omega}}\right]^{1/2}. \tag{9.8}$$

Up to this point, we have assumed that the sensor is not operating in the background limit. For background-limited search systems, we can use Eqs. (8.23) and (8.29) to give the spectral D^* expression of Eq. (9.9):

$$D^*_{\text{BLIP}}(\lambda, T_{bkg}) \approx \frac{\lambda}{hc}\sqrt{\frac{2\eta}{\pi}}\frac{F/\#}{\sqrt{\int_0^{\lambda_{cut}} L_{q,\,bkg}(\lambda, T_{bkg})d\lambda}}. \tag{9.9}$$

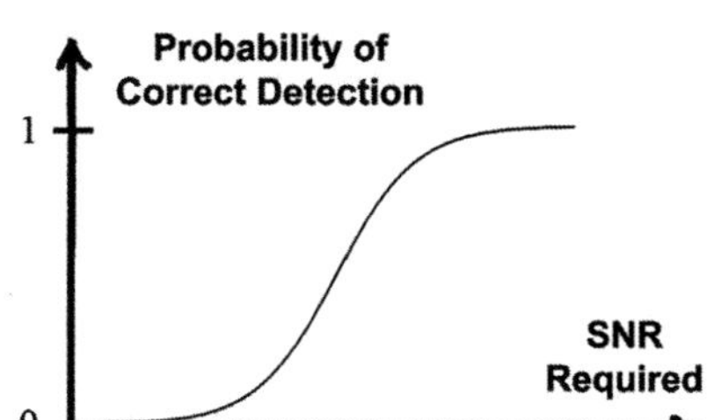

Figure 9.9 Probability of correct detection vs SNR required.

Substituting Eq. (9.9) into Eq. (9.8) gives a standoff range for a BLIP search system that is independent of the optics *F/#* and proportional to the inverse fourth root of the integrated background radiance.

9.3.1 Example of a search system preliminary design

As an example of the design procedure, we will consider a system with the following specifications and assumptions:

- Single detector, two-dimensional raster scan, unit scan efficiency;
- D^* of 1×10^{10} cm $\sqrt{\text{Hz}}/\text{W}$, averaged over the system's spectral passband, which is centered on an operating wavelength of 10 μm;
- Sensor not operating in the background limit;
- Detector dimensions 50 μm × 50 μm;
- Collection optic $F/\# = 2$, $D = 10$ cm, $f = 20$ cm;
- Required SNR = 2.5 for desired detection & false-alarm probability;
- Field coverage required: HFOV × VFOV = 1.8° × 1.8°;
- Target is a point source at a distance of 10 km;
- Target has an in-band intensity of 3 W/sr; and
- Atmospheric attenuation is to be neglected.

We will investigate the following design issues:

- Find maximum frame rate consistent with above specs.
- If sensor is changed to a 128 × 1 linear array (pushbroom scan, with same D^*), what is new maximum frame rate?
- Are there other tradeoffs that can be made?

We will follow these design procedures:

- Calculate flux on detector ϕ from radiometry.
- Using the equation for D^*, knowing sensor area & required SNR, solve for bandwidth.
- Use the equation $\Delta f = 1/(2\tau_d)$ relating bandwidth and dwell time.
- Given IFOV and field-coverage requirements, solve for the frame rate.

This procedure is equivalent to using Eq. (9.8), but our step-by-step approach allows more insight into the tradeoffs involved.

9.3.1.1 Finding the maximum frame rate consistent with initial specs

First, we calculate the flux collected by the lens from the point source:

$$\phi_{\text{det}} = I \times \Omega_{\text{lens}} = 3\,\text{W/sr} \times \frac{\pi}{4}\left[\frac{10 \times 10^{-2}\ \text{m}}{10 \times 10^{3}\ \text{m}}\right]^2 = 2.4 \times 10^{-10}\ \text{W}. \quad (9.10)$$

We then use the expression for D^* in Eq. (8.5) and solve for the bandwidth:

$$\Delta f = \left(\frac{D^* \phi_{\text{det}}}{\text{SNR}\sqrt{A_{\text{det}}}}\right)^2 = \left(\frac{10^{10} \times 2.4 \times 10^{-10}}{2.5 \times 50 \times 10^{-4}}\right)^2 = 36.9 \text{ kHz}. \tag{9.11}$$

Using the equation $\Delta f = 1/(2\tau_d)$, we solve for the dwell time:

$$\tau_d = 1/(2\Delta f) = (2 \times 36.9 \text{ kHz})^{-1} = 13.6 \ \mu\text{sec}. \tag{9.12}$$

Because we know from the problem statement that

$$\text{HIFOV} = \sqrt{A_{\text{det}}}/f = (50 \times 10^{-6} \text{ m})/(20 \times 10^{-2} \text{ m}) = 250 \ \mu\text{rad}, \tag{9.13}$$

we can solve for the scan velocity using the dwell time and Eq. (9.1):

$$\mathcal{V}_{\text{scan}} = \text{HIFOV}/\tau_d = 250 \ \mu\text{rad}/13.6 \ \mu\text{sec} = 18.4 \text{ rad/sec}. \tag{9.14}$$

Using Eq. (9.2), the scan velocity, and the HFOV, we can solve for the time per line:

$$T_{\text{line}} = \text{HFOV}/\mathcal{V}_{\text{scan}} = 250 \ \mu\text{rad}/(18.4 \text{ rad/sec}) = 1.7 \text{ msec}. \tag{9.15}$$

From the problem statement, the $1.8° \times 1.8°$ $(31.9 \text{ mrad})^2$ total FOV is 128^2 IFOVs so that $N_{\text{lines}} = 128$. Using Eq. (9.2), we solve for the frame time:

$$T_{\text{frame}} = N_{\text{lines}} \times T_{\text{line}} = \frac{\text{VFOV}}{\text{VIFOV}} \times \frac{\text{HFOV}}{\mathcal{V}_{\text{scan}}} = 128 \times 1.7 \text{ msec} = 0.22 \text{ sec}. \tag{9.16}$$

The frame rate is the inverse of the frame time, which is = 4.6 frames/sec. This is as fast as the system can operate and still meet the SNR requirements. A slower frame rate will give better SNR, while a faster frame rate will decrease SNR. The rate of 4.6 frames/sec is typically too slow for most applications (standard video is 30 frames/sec), so let us try a different sensor configuration to increase the frame rate.

9.3.1.2 Finding the maximum frame rate using a one-dimensional detector array

We want to determine the new maximum frame rate if the single sensor is changed to a 128×1 linear array with the same D^* operating in a pushbroom scan. Having more sensors allows us to increase the frame rate while keeping the SNR. In the pushbroom-scan format, each line has its own dedicated sensor. Keeping the dwell time at 13.6 μs (consistent with SNR = 2.5) yields a frame time equal to the previous line time of 1.7 ms. Thus, the system can form frames 128 times faster than before, at $128 \times 4.6 = 582$ frames/sec. This frame rate is faster than is needed for most applications.

9.3.1.3 Using excess SNR to increase the maximum range

We now keep the 128×1 sensor format but constrain the system to operate at 30 frames/sec. The resulting longer dwell time will give us an SNR higher than the required value of 2.5, which we can trade off to either increase the range or decrease the front-end diameter of the optics. A frame rate of 30 frames/sec yields a frame time of 33.3 ms. For a pushbroom scan, 33.3 ms is also the line time, given as $128 \times \tau_d$. This gives a dwell time $\tau_d = 258$ μsec. Using $\Delta f = 1/(2\tau_d)$ yields $\Delta f = 1.9$ kHz. The SNR increases to $\mathrm{SNR} = \frac{D^* \phi_d}{\sqrt{A_{\mathrm{det}}}\sqrt{\Delta f}} = \frac{10^{10} \times 2.4 \times 10^{-10}}{50 \times 10^{-4}\sqrt{1900}} = 11$. Our design went from an SNR of 2.5 at 582 frames/sec ($\Delta f = 37$ kHz) to an SNR of 11 at 30 frames/sec ($\Delta f = 1.9$ kHz). These results are consistent with $\mathrm{SNR} \propto 1/\sqrt{\Delta f}$, as seen in Eq. (9.6). At 30 frames/sec, the system achieves $\mathrm{SNR} = 11$. If all we really need is $\mathrm{SNR} = 2.5$, we can increase the range farther than the original value of 10 km, given that we have a factor of (11/2.5) excess ϕ_{det}. The required flux for $\mathrm{SNR} = 2.5$ is $\phi_{\mathrm{det}} = 2.4 \times 10^{-10} \times (2.5/11) = 5.5 \times 10^{-11}$ W. Referring to Eq. (3.6), we see that $\phi_{\mathrm{det}} = I \times \Omega_{\mathrm{lens}} = 3\mathrm{W/sr} \times (\pi/4)[10\,\mathrm{cm}/R_{\mathrm{new}}\,\mathrm{m}]^2 = 5.5 \times 10^{-11}$ W, which gives a new maximum range $R = 20.7$ km. This is consistent with the range dependence $R \propto 1/\sqrt[4]{\Delta f}$ seen in Eq. (9.8).

9.3.1.4 Using excess SNR to decrease the front-end size

If the original 10-km range was sufficient, then the extra SNR can be used to decrease the front-end diameter of the optics. The required flux on the detector for an SNR of 2.5 is $\phi_{\mathrm{det}} = 5.5 \times 10^{-11}$ W. Going back to the radiometry, we see that $\phi_{\mathrm{det}} = I \times \Omega_{\mathrm{lens}} = 3\,\mathrm{W/sr} \times \frac{\pi}{4}\left[\frac{D_{\mathrm{new}}}{10\ \mathrm{km}}\right]^2 = 5.5 \times 10^{-11}$ W. We can solve for $D_{\mathrm{new}} = 4.8$ cm, decreased from 10 cm. A factor of ≈ 4 in signal power is equivalent to a factor of ≈ 2 in lens diameter. A 4.8-cm diameter is sufficient from a flux-transfer viewpoint, but we should also consider the impact of diffraction on image quality. We now have an *F*/4 system rather than the previous *F*/2. At $\lambda = 10$ μm, the *F*/2 system gives a diffraction spot diameter of around 50 μm, about the same dimension as the sensors. But, for the smaller aperture diameter, the diffraction spot diameter is around 100 μm, which will negatively affect image quality.

9.4 Thermal Imager: NETD Expression

A system whose function is to produce a thermal image has a different function compared to a search system. An extended source is to be imaged, and hence radiance (rather than intensity) is the variable of interest, as seen in Fig. 9.10. The irradiance at the image plane is independent of range if an extended source fills the FOV. Temperature differences are inferred from differences in radiant flux emitted, which is a nonlinear function of temperature.

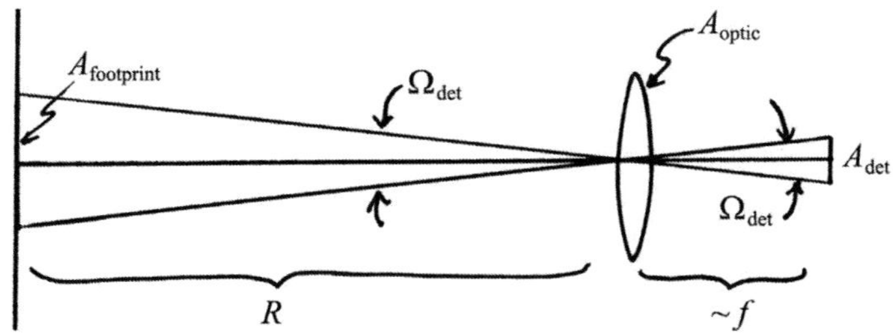

Figure 9.10 Thermal imager configuration.

From radiometry, we find that

$$\phi_{det} = L \times A_{optic} \times \Omega_{det}. \tag{9.17}$$

The solid-angle IFOV can be found as

$$\Omega_{det} = \frac{A_{det}}{f^2} = \frac{A_{footprint}}{R^2}, \tag{9.18}$$

which gives the flux reaching the detector as

$$\phi_{det} = L \times \frac{\pi}{4} D_{optic}^2 \times \frac{A_{det}}{f^2} = \frac{\pi}{4} L \times \frac{A_{det}}{(F/\#)^2}. \tag{9.19}$$

Relating this flux on the detector to a signal voltage v_{sig} by means of responsivity yields

$$v_{sig} = \mathcal{R} \times \frac{\pi}{4} L \times \frac{A_{det}}{(F/\#)^2}. \tag{9.20}$$

Because we are interested in a thermal mapping of the object scene, the change in signal voltage for a change in temperature is of interest:

$$\frac{\Delta v_{sig}}{\Delta T} = \mathcal{R} \times \frac{\pi}{4} \times \frac{\partial L}{\partial T} \times \frac{A_{det}}{(F/\#)^2}. \tag{9.21}$$

Expressing responsivity in terms of sensor D^*: $\mathcal{R} = \frac{\sigma_v \times D^*}{\sqrt{A_{det}}\sqrt{\Delta f}}$, we find that

$$\frac{\Delta v_{sig}}{\Delta T} = \frac{\sigma_v \times D^*}{\sqrt{A_{det}}\sqrt{\Delta f}} \times \frac{\pi}{4} \times \frac{\partial L}{\partial T} \times \frac{A_{det}}{(F/\#)^2}. \tag{9.22}$$

We can rearrange Eq. (9.22) in terms of SNR:

$$\frac{\Delta v_{sig}}{\sigma_v} = \Delta T \frac{D^*}{\sqrt{A_{det}}\sqrt{\Delta f}} \times \frac{\pi}{4} \times \frac{\partial L}{\partial T} \times \frac{A_{det}}{(F/\#)^2}. \tag{9.23}$$

The noise-equivalent temperature difference (NETD) is that temperature difference that gives SNR = 1:

$$\text{NETD} = \frac{\Delta T}{\text{SNR}} = \frac{\Delta T}{\Delta v_{\text{sig}}/\sigma_v}. \tag{9.24}$$

Setting SNR = 1 and ΔT = NETD in Eq. (9.23) yields

$$1 = \text{NETD} \times \frac{D^*}{\sqrt{A_{\text{det}}}\sqrt{\Delta f}} \times \frac{\pi}{4} \times \frac{\partial L}{\partial T} \times \frac{A_{\text{det}}}{(F/\#)^2} \tag{9.25}$$

and

$$\text{NETD} = \left(\frac{4}{\pi}\right) \frac{(F/\#)^2\sqrt{\Delta f}}{\frac{\partial L}{\partial T} \times D^* \times \sqrt{A_{\text{det}}}}. \tag{9.26}$$

The NETD is a measure of thermal sensitivity for an imaging system, indicating how small of a temperature difference we can see in the image. For instance, if a ΔT = 1 K gives SNR of 5, then NETD = 200 mK. Like all noise-equivalent quantities, smaller numbers indicate better performance. We note from Eq. (9.26) that NETD gets smaller for bigger A_{det}. This dependence is why NETD is not used as a design criterion. A larger detector collects more flux, leading to a higher SNR and better thermal sensitivity. However, a larger detector footprint will reduce the image quality, as seen in Fig. 2.11, because of the integration of irradiance over the pixel area. Thus, although NETD does not account for image-quality issues, it is easy to measure. As seen in Fig. 9.11, NETD is a useful day-to-day laboratory test to confirm that the electronics subsystem of an imager is in good working order.

The quantity $\partial L/\partial T \times D^*$ in Eq. (9.26) is an average value, integrated over the spectral passband of the system. For the smallest (best) NETD, the $D^*(\lambda)$ of the sensor should be peaked near the wavelength of maximum

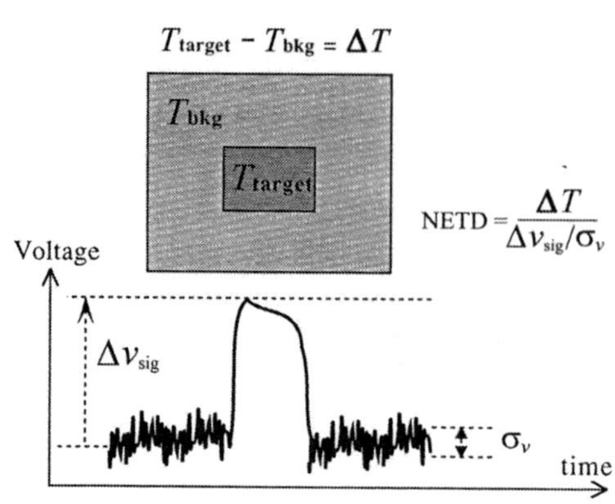

Figure 9.11 NETD test pattern and resulting output voltage waveform.

exitance contrast. Equation (9.26) shows that NETD is proportional to the $(F/\#)^2$ of the imaging system. A smaller *F*/# gives more irradiance at the detector plane according to Eq. (3.26), which allows the collection of more flux from the object. This gives a more accurate estimate of scene temperature. A larger optics aperture collects more flux, and a shorter focal length increases the detector footprint in the scene, again collecting more flux. NETD is proportional to the square root of the signal bandwidth, so a narrower bandwidth (slower frame rate) reduces the noise. If the system is operating in background-limited (BLIP) conditions, then D^* is proportional to *F*/# per Eq. (8.29), which then makes NETD proportional to *F*/# rather than the square of *F*/#.

9.4.1 Calculation example for NETD

Infrared detectors are specified by their D^* values, whereas thermal imagers are typically specified by the system's NETD values. It is of interest to be able to convert between the two. For this example, we assume a staring uncooled IR imager operating in the 8- to 12-μm band, with 50-μm detectors, *F*/1 optics, and a 30-Hz bandwidth. If the NETD spec is 50 mK, what sensor D^* does this correspond to?

First, we evaluate the change in radiance per degree K (as seen in Fig. 4.10) and integrate over the system's spectral passband:

$$\frac{\partial L}{\partial T} = \int_{8\mu\text{m}}^{12\mu\text{m}} \frac{\partial L(\lambda, T = 300\,\text{K})}{\partial T} \text{d}\lambda \approx 6.4 \times 10^{-5} \frac{\text{W}}{\text{cm}^2\,\text{sr}\,\text{K}}. \tag{9.27}$$

Substituting this value into Eq. (9.26), the corresponding integrated in-band D^* value is $\approx 4 \times 10^8$ cm $\text{Hz}^{1/2}$ W^{-1}.

9.5 Minimum Resolvable Temperature Difference (MRTD)

Because NETD characterizes thermal sensitivity at the expense of spatial resolution, the minimum resolvable temperature difference (MRTD) was developed as a figure of merit for the design of thermal imager systems, which combines thermal sensitivity and spatial resolution. MRTD answers the question of what temperature difference is required for bar targets of different (fundamental) spatial frequencies to be visible on a display. Smaller bar targets will require a higher temperature difference because fine details are filtered out by the system's modulation transfer function (MTF). MRTD is measured as shown in Fig. 9.12.

The binary transmissive bar targets are backlit by an extended blackbody source. The temperature difference ΔT between the blackbody and the target plate needs to be controlled to about a factor of ten more precisely than the expected NETD of the system, usually verified by

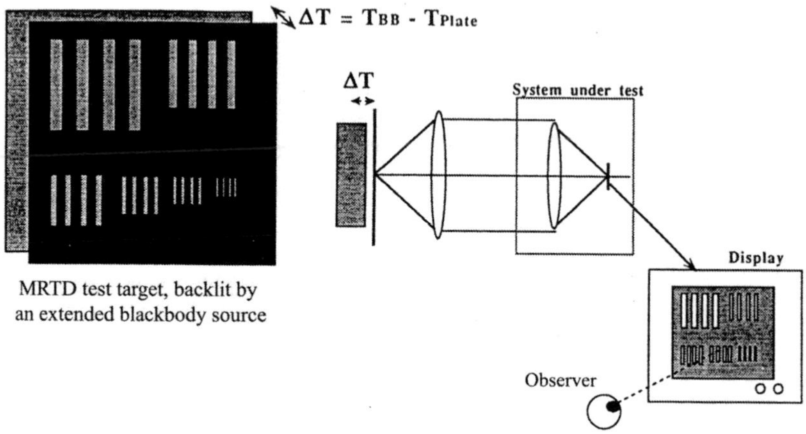

Figure 9.12 MRTD measurement configuration.

thermocouple measurements. The emissivity of the target plate should be uniform and near unity. The bar targets are placed at the focus of a collimator (usually an off-axis parabola) to simulate the target-at-infinity condition that the imager was designed for. The collimator image quality must be sufficient not to affect the measurement at the spatial frequencies (inverse of bar spacings) of interest. Begin with the lowest-frequency target and start with $\Delta T = 0$. Then raise the ΔT until the bars are confidently seen by an observer looking at the display. The display gain should be set so that the system noise is just visible. Repeat this process with successively smaller bars until they cannot be resolved at any temperature difference. These measurements will generate a plot as seen in Fig. 9.13.

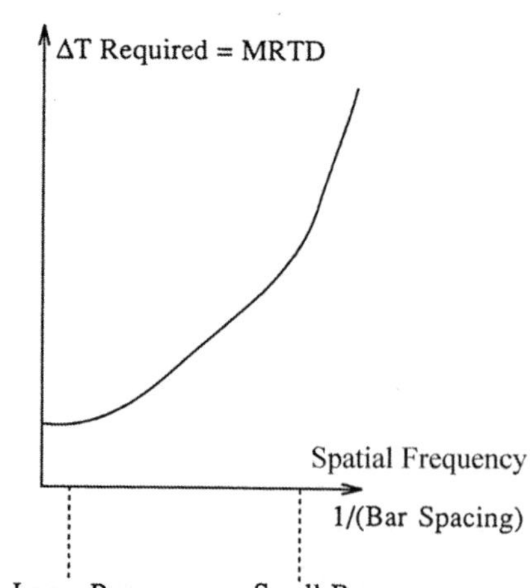

Figure 9.13 Typical MRTD plot.

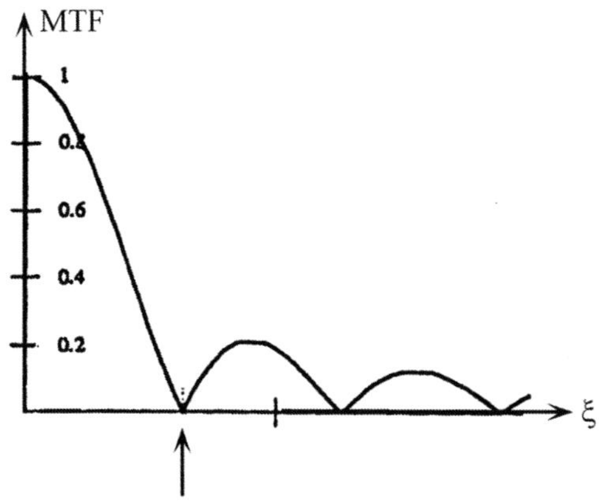

Figure 9.14 MTF of the detector footprint.

The MRTD has a vertical asymptote at the spatial frequency where the MTF first goes through a zero. One component of the MTF is the detector footprint MTF, Eq. (2.11), which is a sinc function (shown in Fig. 9.14):

$$\mathrm{MTF}_{\mathrm{footprint}}(\xi) = \left| \frac{\sin(\pi\ \xi\ \mathrm{IFOV})}{(\pi\ \xi\ \mathrm{IFOV})} \right|. \tag{9.28}$$

The footprint MTF has its first zero at $\xi = 1/\mathrm{IFOV}$. The MRTD will be infinite at that spatial frequency because $\xi = 1/\mathrm{IFOV}$ is completely filtered out by the detector-footprint MTF. MRTD is useful as a performance descriptor and design criterion for IR imaging systems because it accounts for both spatial resolution and thermal sensitivity. It correlates well with target recognition in electronic-noise-limited imagery but does not include effects of scene clutter. One disadvantage of MRTD is that it involves a human observer, which is time-consuming and also introduces variability into the measurement. There has long been a desire on the part of the community for an "objective MRTD" measurement procedure, which would involve separate measurements of MTF and thermal sensitivity. However, a consensus on specifics of how to convert these measurements into MRTD has thus far been elusive.

References

1. T. Tajime, T. Saheki, M. Kondo, and K. Ito, "Serial scan thermal imaging system," *Proc. SPIE* **0156**, 241–245 (1978) [doi: 10.1117/12.956784].
2. http://modtran.spectral.com/modtran_index
3. R. Shirkey and L. Duncan, "The Electro-Optical Systems Atmospheric Effects Library," *Proc. SPIE* **302**, 19–23 (1982) [doi: 10.1117/12.932624].

Bibliography

Daniels, A., *Field Guide to Infrared Systems, Detectors, and FPAs*, *3rd Edition*, SPIE Press, Bellingham, Washington (2018) [doi: 10.1117/3.2315935].

Dereniak, E. L., and G. D. Boreman, *Infrared Detectors & Systems*, Wiley, New York (1996).

Dudzik, M., Ed., *The Infrared & Electro-Optical Systems Handbook, Vol. 4: Electro-Optical Systems Design, Analysis, and Testing*, SPIE Press, Bellingham, Washington (1993) [doi: 10.1117/3.2543823].

Hudson, R. D., Jr., *Infrared Systems Engineering*, Wiley, New York (2006).

Lloyd, J. M., *Thermal Imaging Systems*, Plenum, New York (1975).

Seyrafi, K., *Electro-Optical Systems Analysis*, Electro-Optical Research Company, Los Angeles (1985).

Spiro, I. J., and M. Schlessinger, *Infrared Technology Fundamentals*, Marcel Dekker, New York (1989).

Vollmer, M., and K.-P. Möllmann, *Infrared Thermal Imaging: Fundamentals, Research and Applications*, *2nd Edition*, Wiley VCH, Weinheim, Germany (2018).

Index

$1/f$ noise, 124

A
aberrations, 16
ac coupling, 124
amplifier noise, 107
aperture stop, 11, 46
area solid-angle ($A\Omega$) product, 35

B
background-limited (BLIP) performance, 133
bandwidth, 145, 150
bias current, 83
blackbody, 52
blackbody responsivity, 76
bolometer, 83
brightness temperature, 101

C
carrier lifetime, 92
cavity radiation, 52
chief ray, 13, 47
cold shield, 38, 120, 136
color temperature, 103
cooling, 78
cutoff wavelength, 78, 90

D
dark current, 79, 109
detectors, 67
diffraction, 14, 49, 152
dwell time, 145, 151

E
electron mobility, 91
emissivity, 60, 99
energy gap, 72
energy-derived units, 73
entrance pupil, 46
exit pupil, 46
exitance, 29, 39
exitance contrast, 60
extrinsic photoconductor, 89

F
F/#, 11, 45
feedback resistance, 86
feedback resistor, 87
field of view, 12
focal length, 8
footprint, 13, 143, 157
frame rate, 106, 151
frame time, 145

G
generation–recombination noise, 121
graybody, 62, 101

I
image plane, 10
impulse response, 19, 48
in-band flux, 58
instantaneous field of view, 13, 143
intensity, 29, 31, 41

intrinsic photoconductor, 89
irradiance, 29

J
Johnson noise, 122, 137

K
K-factor, 77
Kirchhoff's law, 53, 61, 120

L
Lambertian radiator, 36
Lambertian source, 33, 41
long-wave infrared, 17
long-wavelength cutoff, 72

M
magnification, 10
marginal ray, 11
maximum range, 149
measurement
    bandwidth, 109, 129
mid-wave infrared, 17
minimum resolvable temperature
    difference, 155
modulation depth, 23
modulation transfer
    function, 19, 155

N
near infrared, 17
noise, 107
noise-equivalent
    bandwidth, 79
noise-equivalent power, 127
noise-equivalent temperature
    difference, 154
normalized detectivity D^*, 129

O
open circuit, 94
optical axis, 8

P
parallel scan, 146
peak-spectral D^*, 134
photoconductive detector, 89
photoconductive gain, 92, 121
photon detector, 67, 89
photon-derived units, 73
photovoltaic detector, 92
Planck's equation, 54, 59
p–n junction, 92, 114
point source, 48, 148
power spectral density, 110
pushbroom scan, 146
pyroelectric sensor, 85

Q
quadrature, 107, 147
quantum efficiency, 70, 117

R
RA product, 138
radiance, 29, 33, 47
radiation temperature, 101
responsivity, 67, 69, 130
reverse bias, 95
reverse-saturation current, 95

S
scan velocity, 145
scanning, 143
Schottky-barrier detector, 96
search system, 143, 147
selective radiator, 63
sensitivity, 67
serial scan, 146
short circuit, 94
short-wave infrared, 17
shot noise, 79, 113
signal-shot-noise limit, 116
signal-to-noise ratio, 67
spatial frequencies, 20
spectral exitance, 51

spectral flux, 51
spectral radiance, 51
spectral responsivity, 70, 73, 74
spectral transmittance, 17
spectroradiometer, 104
Stefan–Boltzmann law, 54, 55

T
thermal capacitance, 81
thermal conductance, 81
thermal detector, 67, 75, 81
thermal fluctuation, 139
thermal-fluctuation noise, 125
thermal-imager system, 143
throughput, 36
time delay and integration, 146

U
uncorrelated waveforms, 107

W
white noise, 112
Wien's displacement law, 54, 56

Glenn D. Boreman is Professor and Chairman of the Department of Physics & Optical Science and Director of the Center for Optoelectronics & Optical Communications at the University of North Carolina at Charlotte. He is co-founder and Board Chairman of Plasmonics, Inc. (Orlando). From 1984 to 2011 he was on the faculty of the University of Central Florida, where he is now Professor Emeritus. He has supervised to completion 35 MS and 27 PhD students. He has held visiting research positions at IT&T (Roanoke), Texas Instruments (Dallas), US Army Night Vision Lab (Ft. Belvoir), McDonnell Douglas Astronautics (Titusville), US Army Redstone Arsenal (Huntsville), Imperial College (London), Universidad Complutense (Madrid), Swiss Federal Institute of Technology (Zürich), Swedish Defense Research Agency (Linköping), and University of New Mexico (Albuquerque). He received the BS in Optics from the University of Rochester, and the PhD in Optics from the University of Arizona. Prof. Boreman served as Editor-in-Chief of *Applied Optics* from 2000 to 2005, and Deputy Editor of *Optics Express* from 2014 to 2019. He is coauthor of the graduate textbooks *Infrared Detectors and Systems* and *Infrared Antennas and Resonant Structures* (SPIE Press), and author of *Modulation Transfer Function in Optical & Electro-Optical Systems* (first and second editions, SPIE Press) and *Basic Electro-Optics for Electrical Engineers* (SPIE Press). He has published more than 200 refereed journal articles in the areas of infrared sensors and materials, optics of random media, and image-quality assessment. He is a fellow of SPIE, IEEE, Optica, and the Military Sensing Symposium. He is a Professional Engineer registered in Florida. Prof. Boreman served as the 2017 President of SPIE.